Nuclear Remains

Inside Technology

Edited by Wiebe E. Bijker, Edward Jones-Imhotep, and Rebecca Slayton

A list of books in the series appears at the back of the book

Nuclear Remains

On Temporalities, Responsibilities, and Values

Başak Saraç-Lesavre

The MIT Press
Cambridge, Massachusetts
London, England

The MIT Press
Massachusetts Institute of Technology
77 Massachusetts Avenue
Cambridge, MA 02139
mitpress.mit.edu

The MIT Press would like to thank the anonymous peer reviewers who provided comments on drafts of this book. The generous work of academic experts is essential for establishing the authority and quality of our publications. We acknowledge with gratitude the contributions of these otherwise uncredited readers.

This book was set in Stone Serif and Stone Sans by Westchester Publishing Services. Printed and bound in the United States of America.

Library of Congress Cataloging-in-Publication Data

Names: Saraç-Lesavre, Başak author
Title: Nuclear remains : on temporalities, responsibilities, and values / Başak Saraç-Lesavre.
Description: Cambridge, Massachusetts : The MIT Press, [2026] | Series: Inside technology | Includes bibliographical references and index.
Identifiers: LCCN 2025028955 (print) | LCCN 2025028956 (ebook) | ISBN 9780262052856 paperback | ISBN 9780262052863 pdf | ISBN 9780262052870 epub
Subjects: LCSH: Radioactive waste disposal—United States—History | Radioactive waste disposal—Government policy—United States
Classification: LCC TD898.118 .S27 2026 (print) | LCC TD898.118 (ebook)
LC record available at https://lccn.loc.gov/2025028955
LC ebook record available at https://lccn.loc.gov/2025028956

EU Authorised Representative: Easy Access System Europe, Mustamäe tee 50, 10621 Tallinn, Estonia | Email: gpsr.requests@easproject.com

À Julien

Contents

Acknowledgments

First and foremost, I am indebted to all the concerned actors from whom I learned tremendously through both formal interviews and hundreds of hours of informal discussions. Their generosity, openness, and grace have been extraordinary. Some even welcomed a "visitor alien" into their homes and families without hesitation. If I maintain faith in humanity, it is, in part, thanks to them. I owe gratitude to Abe Van Luik; I wish you were still with us to witness this book's completion, whether you agreed with its arguments or not. Allison Macfarlane's instrumental support has been foundational to this research. Her exceptional expertise, profound knowledge of the nuclear field, and unwavering advocacy opened countless doors that would have otherwise remained closed, fundamentally shaping the trajectory and depth of this project.

Special thanks to Peter Galison, whose intellectual generosity significantly shaped this project. Beyond providing an institutional affiliation at Harvard University's Department of History of Science, his insightful feedback, thoughtful mentorship, and belief in this work's importance were crucial in the development of my ideas. I am grateful to the Franco-American Fulbright Commission for generously funding and facilitating my fieldwork and to the Centre National de la Recherche Scientifique for financing my research project.

This project was shaped early on by intellectual guidance from the Center for Sociology of Innovation at the Ecole des Mines de Paris. Fabian Muniesa consistently encouraged me to direct my attention to details overlooked by most and has provided me with outstanding intellectual mentorship. Alexandre Mallard, Antoine Hennion, Vololona Rabeharisoa, Madeleine Akrich, Brice Laurent, Morgan Meyer, and David Pontille posed challenging questions, collectively teaching me to be fascinated by "what actors do" rather than "what they say they do" and urging me "to follow them in action." I took their counsel quite literally, regularly revisiting their insights as I worked through countless drafts.

Penny Harvey's kindness, patience, and profound intellectual curiosity expanded my thinking across time, regions, subfields, disciplines, and theoretical paradigms. For five years, colleagues at the University of Manchester's Department of Social Anthropology bolstered this work through their wisdom, expertise, and collegiality. Our engaging departmental retreats, seminars, and reading and writing groups proved instrumental in refining my writing and theoretical ideas.

One of my extended discussions on the problem of rendez-vous concept occurred at an intimate writing group organized by my exceptional colleague Caroline Parker. Chika Watanabe, Sebastien Bachelet, Soumhya Venkatesan, and Caroline Parker found the concept compelling and urged me to develop it further. Karen Sykes consistently shared her tremendous knowledge and wisdom on intergenerational relations with remarkable generosity. I remain indebted to Caroline Parker for her unwavering support, incisive questions, and illuminating feedback throughout various iterations of this manuscript, which proved crucial to its completion. Additional appreciation goes to Peter Wade, Constance Smith, Chloé Nahum-Claudel, Sebastian Bachelet, Michelle Obeid, Rupert Cox, Jolynna Sinanan, Tony Simpson, and Madeleine Reeves for their intellectual curiosity and camaraderie. The Beam, the University of Manchester's Social and Nuclear Research Network, has hosted my research for several years. I am particularly thankful to Penny Harvey, Richard Taylor, Petra Kalshoven, Damien O'Doherty, and Ian Tellam for our deeply stimulating exchanges on ethnography, infrastructures, and nuclear matters.

I was fortunate to spend a year as a postdoctoral fellow at Virginia Tech's Department of Science, Technology and Society, which offered an extremely stimulating research and teaching environment and an excellent opportunity to pursue further different aspects of this research. I especially thank Sonja Schmid for her invaluable guidance in all dimensions of my work and for her steadfast mentorship and unwavering support. I also thank my wonderful colleagues Daniel Breslau and Gary Downey for providing critical feedback that helped advance my ideas. That year, nuclear seminar series hosted by Allison Macfarlane and Hugh Gusterson at George

Washington University's Elliott School of International Affairs, proved thought-provoking for my research.

I had the privilege of presenting elements of this book to numerous audiences who provided opportunities to test my ideas and challenged me to refine my arguments. These include audiences at the University of Manchester's Department of Anthropology; University of Vienna's Department of Science and Technology Studies; Virginia Tech's Department of Science, Technology and Society; University of Saint Andrew's Department of Social Anthropology; University of Edinburgh's Department of Science, Technology and Innovation Studies; University College London's Department of Geography; Harvard University's Department of the History of Science; Ecole des Hautes Etudes de Sciences Sociales' Centre Alexandre Koyré; and Ecole des Hautes Etudes de Sciences Sociales' Centre d'Etudes de Mouvements Sociaux. Elements of this book also appear in the *Journal of Cultural Economy* and *Environment, Planning D: Society and Space*.

My first journey to the United States was for the STGlobal Conference organized by the American Association for the Advancement of Science during my graduate program's first year. Of all possible academics, Daniel Sarewitz and Daniel Metlay served as discussants on my panel. They could have responded in countless ways to such a preliminary paper on US nuclear waste policy presented by a foreigner, yet they chose to offer encouragement with remarkable generosity and provide invaluable advice.

My research recently found a new home at Sciences Po's Centre de Sociologie des Organisations. I had the pleasure of presenting the revised manuscript at our Department's Savoirs, Sciences et Expertise seminar series. I am thankful to all participants, particularly Jean-Noël Jouzel, Renaud Crespin, Sylvain Brunier, Manisha Anantharaman, Damien Garcia, and Reiko Hasegawa for their stimulating and thought-provoking remarks that significantly aided my final revisions. I would also love to thank other colleagues for their curiosity and generous support, including but not limited to, Sophie Dubuisson-Quellier, Olivier Pilmis, Kevin Mellet, Sarah Gensberger, Jérôme Aust, Olivier Borraz, Patrick Castel, Marie-Emmanuelle Chessel, Martin Giraudeau, Jeanne Lazarus, Etienne Nouguez, Gwenaëlle Rot.

I am deeply appreciative of Rodney Ewing, Thomas Isaacs, Mary Woollen, and Per Peterson for critically engaging with my opinion piece on the current state of US nuclear waste policy published in *Issues in Science and Technology*. Throughout the editorial process, Daniel Sarewitz's intellectual and editorial engagement with my piece constituted an enriching intellectual conversation that provided extremely valuable insights for articulating further my findings about nuclear waste policy.

Various research workshops, theme-specific conferences, and panel sessions organized by numerous outstanding colleagues helped me test my ideas with scholarly audiences sharing similar research interests. Among these, I acknowledge feedback received on different arguments at "Le sol, nouvelle frontière pour les sciences et les politiques de l'environnement" by Céline Granjou and Germain Meulemans at EHESS; "The Nuclear and Social Science Nexus: Challenges and Opportunities for Speaking across the Disciplinary Divide," organized by Aditi Verma, Sama Bilbao Y. Leon, and Markku Lehtonen at the Organisation for Economic Co-operation and Development (OECD); the Discard Studies Conference on "Exploring Disposal's Past, Present, and Future" at NYU Gallatin Study for Individualized Studies, organized by Rosalind Fredericks, Mohammed Rafi Arefin, and Robin Nagle; 4S Conference panel session, organized by Hugh Gusterson and Allison Macfarlane; "Negotiating the Nuclear Divide: The Case of the Classification of High-Level Waste in the United States," organized by Taylor Loy and Sonja Schmid at Virginia Tech; and "Penser le Soin à travers ses Tensions: Fragilité et Force du Vivant et de Son Milieu," organized by Laurence Tessier at EHESS's Centre d'Étude des Mouvements Sociaux, Institut Marcel Mauss. The research workshop on "Containers and Waste: Ordering the Unruly," organized by Daniel Sosna and David Hanig, provided an outstanding opportunity to test my ideas among scholars working on containment; I am also grateful to them and to Zsuzsa Gille and Sasha Newell for their insightful feedback.

Aditi Verma and Denia Djokovic invited me to different international expert panels they organized at the University of Michigan's Nuclear Science and Engineering Department on the US nuclear policy, offering invaluable opportunities to discuss and articulate my ideas before scholars, experts, and policymakers. Aditi has also proven to be a very supportive friend who consistently asked, with genuine enthusiasm, "when is the book coming out?".

I am deeply appreciative of Wiebe Bijker, Edward Jones-Imhotep, and Rebecca Slayton, my editors at the MIT Press, for their steadfast support of this project. Books from this series have been foundational to my training in Science and Technology Studies, and I have always aspired to publish with this series. I recall my trepidation in submitting my proposal. I sent my initial book proposal to Trevor Pinch, whose immediate and enthusiastic response was reassuring and motivated me to pursue this project. I deeply regret that he will not see this book in print; his pioneering work and generous mentorship continue to inspire countless scholars in the field. I am tremendously thankful to Wiebe Bijker for taking the project forward. I also thank Justin Kehoe and Suraiya Jetha for guiding the book through its final stages. I cannot adequately express my gratitude to the four reviewers who read the manuscript with care and offered detailed, incisive, and constructive feedback. I thank Shanon Fitzpatrick for her invaluable editorial feedback and for having shared my passion for this project. I am also very grateful to Sinéad O'Sullivan for her consistently careful and thorough proofreading, and to all the members of the MIT Press and its partners for their invisible labor in bringing this book to fruition.

Above all, I am grateful for my family. I thank my brother, Umut, for perceiving the world as horizontal. My parents, Nahit and Nuray, fought tirelessly to make the impossible possible for their children. The problem of rendez-vous honors, in part, the major challenges they dealt with over their lifetimes. Julien's immense love, patience, wisdom, insight, without an ounce of doubt, made this book possible.

Introduction

A trip to Yucca Mountain (Nye County, Nevada) must be planned carefully. In the spring of 2012, I was accompanied there by Tom, a retired scientist. We left before sunrise from the town of Summerlin. He was well prepared, sporting a pick-up van with sufficient machinery to climb a mountain, a professional camera for landscape photography, another one to self-film, and a bag containing zooms and lenses. I carried water and my notebook, as if it were any other day at the office. For safety, we had shared our GPS coordinates with at least two people. This was "the desert rule," Tom said. After driving almost two hours, we crossed Amargosa Valley, home to the Shoshone and Paiute peoples and the Nevada Test Site—one of America's most important atomic bomb–testing sites. We were close to our destination, but the usual road to the mountain was closed, so we made the necessary detour and approached the mountain from behind. As we climbed the bumpy path, we passed a Portaloo made of plastic, left abandoned by the side of the road. This defunct repository for waste was a sign of what was to come.

To reach the inaccessible side of the mountain, I needed to climb the last hundred meters on foot. I wanted to see what remained of a deep geological repository project that had once been planned to contain the entirety of the United States' high-level nuclear waste for a million years. As I stepped onto the soil, I picked up a branch from the low-lying, dark grayish-green, and thorny thicket that covered the mountain. Tom said that it was "blackbrush," native to the deserts of the southwestern United States, and suggested I smell it. I inhaled a deep scent of smoke. Hardly a coincidence, as the mountain itself was created by ash falling from volcanic eruptions more than ten million years ago. Over millennia, that ash transformed into "tuff," a type of rock. The crest of Yucca Mountain was formed by the movement of the layers of tuff created as a result of geological forces, and it is crisscrossed by fractures formed by the cooling of the volcanic units. Its geological properties, among many other things, had led the US Congress to designate Yucca Mountain as the sole candidate site to isolate the country's mounting high-level nuclear waste deep inside. Over the ensuing decades, policymakers and institutional experts developed a sophisticated set of technoscientific, financial, legal, economic, material, and political arrangements to produce that containment. Reaching the top of Yucca Mountain, amid an abundance of land and sky, evidence of this project now surrounded me. But all was in a state of arrested development: the worksites emptied of workers, the landscape abandoned. Having lost its funding from Congress, the planned repository—the only one of its kind—was being dismantled.

Over the course of just a few months, I watched as stunned scientists and other employees were let go or reassigned, contracts were terminated, materials were transferred to other federal sites, and equipment (so highly specialized that it was hard to imagine who else would buy it) was put up for sale. Some who had dedicated their careers to the project, like Tom, held out hope that it would be revived; however, many others recommended moving on. For my part, I was left wondering. What did the Yucca Mountain project's abrupt termination mean for the US nuclear waste policy's future and the fate of nuclear waste? And, moreover, what does the story of its demise tell us about our capacities—and incapacities—to ethically address challenges facing humans and the planet? The case of nuclear remains, I realized, dramatizes the vital nature of such questions, revealing how efforts to create "desirable" linkages between the past, present, and future are not only key to achieving a "good" society but perhaps also to having any livable futures at all.

* * *

Nuclear Remains probes how institutions, experts, and communities grapple with the temporal, political, epistemological, and ethical issues that arise when policymakers decide that dealing with the long-lasting remains of existing human activities requires action in the present. Through the case of US nuclear waste,[1] it analyzes how concerned actors attempted to define the terms of "responsible action" and translate these into a whole series of arrangements[2] for taking care of radioactive materials that aimed to establish "good" relations between the past, present, and future. Putting these arrangements into practice, however, turned out to be more difficult than anyone had anticipated or was prepared for. This set into motion complex political debates, financial calculations, economic negotiations, budgetary instruments, and other processes that constituted key sites of valuation and revaluation where temporal relations were continually remade. By unpacking how, when, and where specific arrangements were formulated to collectively produce the physical isolation of nuclear remains deep underground, and by tracing the trajectories they followed over time, this book offers a broader reflection on the challenges involved in producing long-term oriented action.

The US Congress legislated the Nuclear Waste Policy Act of 1982 based on the principle that morally and politically desirable action necessitated the long-term isolation of high-level nuclear waste from humans and the environment for deep time, and it decided that the best way to do this was through geological repositories. This was followed by the adoption of the Nuclear Waste Policy Act Amendments of 1987 that designated Nevada for the building of such a repository project at Yucca Mountain. But after $13.5 billion and thirty-five years of investment, this project was abandoned—also, essentially, through an act (or inaction) of Congress. At the time of writing, there is still no operating geological repository for high-level nuclear waste in the United States or anywhere else in the world. Instead, over 83,000 metric tons[3] of commercial spent nuclear fuel (one form of high-level waste)[4] generated by nuclear power plants is sleeping temporarily across a dispersed network of seventy reactor sites in thirty-four states (Office of Nuclear Energy 2022), while most defense high-level waste is warehoused separately, and equally temporarily, at three federal sites across three states. All these wastes were destined to be buried in the depths of Yucca Mountain, but with the project terminated, how and where they will be handled is once again uncertain, generating a whole range of concerns and debates.

Meanwhile, there still exists, albeit not without controversy, an international agreement that nuclear waste must be safely and securely confined from the biosphere for very long timescales. The International Atomic Energy Agency (IAEA) summarizes this very well. In its *Joint Convention on the Safety of Spent Nuclear Fuel Management and on the Safety of Radioactive Management*, the IAEA states that "effective defenses" should be put in place so that "individuals, society and the environment are protected from harmful effects of ionizing radiation, now and in the future, in such a way that the needs and aspirations of the present generation are met without compromising the ability of future generations to meet their needs and aspirations" (International Atomic Energy Agency 1997). This specific political and moral vision has pushed experts, communities, and institutions to imagine the responsibilities of the present generations toward future ones, drawing attention to the ethical dimensions of the relations that societies build across time with imagined others in the future—who are frequently conceptualized as "future generations."

When establishing specific ethical and political relations between the past, present, and future relies on acting the present, what kind of arrangements render futures present and actionable? And just as importantly, how do related plans and policies *remain* operational and desirable over time? These are clearly very hard questions. But an in-depth analysis of the case of high-level nuclear waste exploring how those issues have been dealt with in the past offers insights into understanding how specific forms of practices, institutions, conventions, technologies, and devices enable—and disable—collective action.

As I am writing these lines, communities across the United States are living and dealing with high-level nuclear waste, while policymakers and governmental institutions are still attempting to reset the national nuclear waste policy. Different actors provide different explanations for the Yucca Mountain project's so-called "failure." Some attribute it to the failure of representative democracy, pointing to the decades-long political opposition the project had faced from the State of Nevada, the Shoshone and Paiute peoples or to the failure of regulatory science to account for various technoscientific controversies over the site's potential incapacity to physically isolate radioactive matter from humans and the environment for long scales of times. Others qualify it as the "failure" of the federal government to take long-term policy action.

Indeed, the Yucca Mountain project isn't operational, and Yucca Mountain will probably never physically contain high-level nuclear waste deep inside. Such pronouncements, however, overlook not only the complexity of arrangements that were to hold the US nuclear waste policy together but also the ways in which nuclear economies, communities, geologies, and geographies have been made to operate and do still operate through the mediation of these arrangements, whether they take the form of contracts, electricity bills, scientific knowledge produced on geological formations, materials or society, annual budgets, regulatory assessments, political agreements, business plans, or management organizations. Through these arrangements, and the tests they underwent, particular forms of US society have been constantly produced—specific ways of organizing temporal relations, distributing responsibilities, and articulating values.

Scholars of science and technology studies have long highlighted that there is much more to learn from "failures" than from "successes," since "failures" render visible the otherwise invisible controversies, negotiations, and compromises that went into formulating all that were to collectively produce a given project (Bijker et al. 1987; Galison 1987; Latour 1992; Bijker 1997). No matter how the Yucca Mountain project should be qualified, I construe an industrial society's efforts to act in the present to deal with the long-lasting remains of its nuclear age as a unique opportunity to tell a story that reshuffles our understanding of a central problem posed by nuclear waste, as well as what the existing attempts to deal with this peculiar material reveal about our collective capacities to establish "good" and "desirable" relations across time.

Problem of Rendez-Vous

Studying the diverse nexus of actors who have attempted to cope with the United States' most hazardous and long-lasting nuclear waste, I have come to understand that the definition of what constitutes responsible action relies on its capacity to resolve (at least partially and temporarily) what I call a *problem of rendez-vous*. This concept grasps situations in which the temporal capacities of existing practices, norms, expertise, institutions, and technologies to mediate relations between the past, present, and future are put to the test. Provoking efforts to identify temporal gaps and to synchronize and coordinate actions between parties who are to act, or even exist, at different moments in time, they prompt concerned actors to think of ways to hold conflicting and competing values, responsibilities, and temporalities together.

In vernacular language, the French expression *rendez-vous* means the deliberate arrangement of a meeting between two or more actors. The meeting is set to bring together actors with all sorts of expectations. Even when the rendez-vous is set with ease, it is of a temporary nature, with the participants negotiating their interests, values, and responsibilities, which often do not neatly overlap. The expression *rendez-vous* takes on an additional meaning in the French language, which makes it worth using here. Beyond meeting at the right time in the right place, being "au rendez-vous" (at the rendez-vous) involves a specific judgment. To be au rendez-vous, it is neither sufficient nor obligatory for all the actors to be present at the right time and in the right place. All the actors may be present at the right time and in the right place, but the conditions of felicity of being au rendez-vous may not be fulfilled or vice versa. For instance, the French gymnastics team was in the right place at the right time to compete at the 2024 Olympics in Paris. However, the French media judged them not to be au rendez-vous at the Olympics, as their performance did not live up to the expectations they had set.

The conditions of felicity of being au rendez-vous might be met even if those who set them up are not physically present at the right time and in the right place. For example, great-grandparents may place a family heirloom in a safe deposit box to ensure that it will be available to their great-grandchildren on a certain date. Great-grandparents may or may not be considered au rendez-vous by their great-grandchildren; these great-grandchildren may never have been born, they may die unexpectedly before the inheritance reaches them, the bank may be robbed, or the inheritance may be materially damaged. The reception of the heirloom by the great-grandchildren at the intended time doesn't mean that they will judge the decision of their great-grandparents to relay the object as good, desirable, or reasonable. This is why the problem of rendez-vous works as an analytical tool for investigating the specific efforts of people, communities, experts, and institutions that try to handle long-lasting responsibilities: which *temporally and spatially defined collectives* are held together? Through what technoscientific, material, financial, legal, political, and moral arrangements? And which are the trajectories that they follow over time? Answering each of these questions, concerned actors attempt to establish good and desirable relations between the past, present, and future and, therefore, to resolve problems of rendez-vous.

While nuclear remains can be construed as a singularly looming danger of a magnitude almost too vast to comprehend, problems of rendez-vous identified in attempts to deal with it are myriad—and indeed sometimes more graspable and therefore actionable. They can range from conflicts over how to contain uncertain financial responsibilities associated with large-scale, long-term projects whose costs may rise or fall over time to competing moral and political notions of a "fair," "accurate," and "efficient" distribution of risks and responsibilities over time. Resolving problems of rendez-vous entails reaching arrangements to contain, in their own specific ways, the concerns and priorities of temporally and spatially defined collectives in desirable manners, at least partially and temporarily. Those arrangements are negotiated not between pregiven generations as such but rather among temporally and spatially defined collectives that come into being as the outcome of various mediations, negotiations, and configurations. The notion of temporally and spatially defined collectives both populates and situates generations in time and place while emphasizing their relational character; it, therefore, enables distancing from preexisting and taken-for-granted temporal categories. Using the notion of temporally and spatially defined collectives instead of generations calls attention to how relations among these collectives are constantly negotiated, established, or dissolved over, through, and by time. Therefore, I concentrate on how temporally and spatially defined collectives resolve problems of rendez-vous, and I unfold how their technologies, values, interests, conceptions of justice, democracy, and expectations about future states of the economy, society and the Earth evolve, along with their very composition, sometimes in ways that have profound implications for the definition of "responsible action."

I further argue that resolutions of problems of rendez-vous—both in regards to nuclear waste but also, as this case will illuminate, for other challenges extending across time—are intimately entangled with political economies in which societies operate. Recently and today, they take on peculiar flavors under the ever-evolving conditions of neoliberalism. As political scientist Will Davies has noted, neoliberal reasoning provides not only an economic vision for society but also a moral one, a moral economy that "seeks to represent all commitments—including those to the future and to future generations—in calculable monetary terms" (Davies 2017, 10). The fate of a nuclear waste disposal facility is not only determined by a political vote in the relevant political institutions but also by a line in a budget or by the choice of an accounting[5] figure selected to estimate the present value of its future costs and benefits. Its operating conditions are enshrined not only in relevant legislation, rules, or regulations but also in the business plans of a company set up to physically manage such waste, a contract signed between a county or city council and a nuclear company, or another contract signed between a nuclear utility and the federal government. In other words, whether in a town in Nevada or New Mexico or at a congressional committee, nuclear waste programs are cluttered with economic, budgetary, and financial logics, as well as their conflicting and competing moralities.

In their attempts to resolve problems of rendez-vous, concerned actors engage in negotiations to hold responsibilities, temporalities, and values together. Resolving problems of rendez-vous includes engaging with future-making practices. Science and technology studies, sociology, and anthropology literatures have long shown how policymakers, scientists, communities, experts, and institutions are constantly brought to imagine future states of society, the economy, and the Earth, so as to make those very futures (e.g., Sarewitz et al. 2000; Masco 2005; Rosenberg et al. 2005; Borup et al. 2006; Jasanoff and Kim 2009, 2015; Beckert 2013, 2016; Pels 2015; Coleman 2017; Bryant and Knight 2019; Beckert and Suckert 2021).

Here, it is useful to situate the problem of rendez-vous among two important concepts that directly grapple with the ways in which futures are produced. One of them is "sociotechnical imaginaries," and the other is "fictional expectations." The notion of sociotechnical imaginaries, developed by Sheila Jasanoff and Sang-Hyun Kim, encourages thinking about imaginaries around which different institutional actors engage in future-making, which are "collectively imagined forms of social life and social order reflected in the design and fulfilment of nation-specific scientific and/or technological projects" (Jasanoff and Kim 2009, 120). They are national, powerful, instrumental, enduring, and futuristic, and need to be "identified, illuminated and critiqued" (Jasanoff and Kim 2009, 120). I contend that paying attention to problems of rendez-vous, instead of examining overarching sociotechnical imaginaries and how they are distilled into specific policy actions, enables, first, tracing situated negotiations, conflicts, and compromises through which communities, experts, and institutions take action to link the past, present, and future and, second, analyzing what this tells us about contemporary societies' capacities (and incapacities) to deal with responsibilities across time.

Second, the notion of fictional expectations, developed by Jens Beckert, is also important to think with. Beckert defines fictional expectations as "images actors form as they consider future states of the world, the way they visualize causal relations and the ways they perceive their actions influencing outcomes" (Beckert 2016, 9). This doesn't mean that they are "false or mere fantasies" but that "actors motivated by an imagined future state, organize their activities based on this mental representation" (Beckert 2016, 9). While studying fictional expectations is important for understanding how concerned actors make the future "present" and "actionable," the notion of the problem of rendez-vous shifts empirical attention to the ways in which they attempt to hold temporally and spatially situated *collectives* together in different forms of arrangements, which are fragile, contested, and constantly subject to tests. In this sense, the problem of rendez-vous framework uniquely (1) shifts the empirical gaze from future-making to the efforts to link the past, present, and future; (2) underlines the ever-evolving, contested, and negotiated nature of such temporal orderings; and (3) allows tracing constant revaluation(s) such arrangements undergo through both their conception and trajectories in action.

This brings me to a critical question that the concept of the problem of rendez-vous helps us grapple with: Where should or must the moral cursor of action be set in time to qualify as "responsible"? To answer this question, let me first think with authors who have dealt with this question in their own ways.

While acknowledging human intervention into deep time, not all scholars agree on how the long-lasting consequences of human activities should be addressed. Sociologists Barbara Adam and Chris Groves refer to future-oriented action as "institutionally constituted irresponsibility" and qualify it as the "politics of posterity" (Adam and Groves 2007). These authors argue that any consideration that involves treating the consequences of modernity in the present are the results of wishful thinking. They consider that when the conventional knowledge system meets the actual outcome of its consequences in time, this meeting can only take the form of a clash that generates paradoxes and unintended consequences, since neither scientific predictions nor control methods can do anything other than fail once they face the contemporary condition.

Whether the Yucca Mountain project specifically or geological repository projects more broadly should be considered as ways to account for the well-being of far futures, or whether they should be understood as sources of irreversible long-lasting environmental problems, is an issue that continuously generates debates and controversies (Shrader-Frechette 1993; Metlay 2000; Oreskes 2000; Masco 2005; Patinaux 2017). In their aptly titled *Acting in an Uncertain World*, sociologists Michel Callon, Pierre Lascoumes, and Yannick Barthe advocate that action should be engineered to remain open, freeing futures of the burden of past decisions, at least politically (Callon et al. 2009). Primarily based on their analysis of the trajectory of the French national geological repository program, the authors advocate for the adoption of an iterative and participatory decision-making process, in parallel with undertaking reversible actions in the present, so that "future generations" would themselves be able to decide whether they qualify the option adopted by "previous generations" as "desirable"—or modify it if they deem it necessary.

Feminist Science and Technology Studies scholars insist that it is time to shift the temporal orientation of action from the future to the present. Philosopher Donna Haraway suggests that "what used to be called nature has erupted into ordinary human affairs, and vice versa, in such a way, and with such permanence as to change fundamentally means and prospects for going on" (Haraway 2016, 40). Under this condition, even if the idea of reopening futures—in other words, undertaking action in the present to avoid futures being colonized by the present—is appealing, anticipatory modes of action are no longer appropriate to protect the Earth, because action can only be "response-able" if those undertaking it remain accountable for its consequences (Tsing 2015; Haraway 2016; Tsing et al. 2017; Weston 2017). However, "response-able" action still holds the potential for paths taken in the present to snowball across generations and to produce a more desirable and valuable future for the Earth and its inhabitants.

A recent stream of work in sociology and political science takes the opposite view and considers that long-term action is desirable and criticizes contemporary societies for failing to take such action. In his book, *How We Sold Our Future?*, sociologist Jens Beckert joins a growing literature that identifies "temporal misalignment between the short-term incentives of politics and the long-term processes of climate change" (Beckert 2024, 69). Political scientist Thomas Hale calls these "long-problems,"[6] defining them as issues with "temporal distance between primary causes and effects" (Hale 2024, 9). Others diagnose "democratic myopia"[7] as a structural flaw preventing adequate responses to challenges extending beyond electoral cycles (Boston 2017, 2021b; MacKenzie 2021; MacKenzie et al. 2023).

These divergent positions reflect fundamentally different approaches about where the moral cursor of action should be set in time. Yet whether scholars advocate future-oriented action, reject future-orientation in favor of present "response-ability," or advocate iterative approaches, I contend that they are all responding to the same analytical challenge: how to establish and maintain "desirable" relations across time amid conflicting, competing, and ever-evolving values and priorities—in other words, how to resolve problems of rendez-vous.

With the notion of problem of rendez-vous, I take distance from such moral premises and transform political and ethical concerns about temporal orientation of action into analytical ones. Rather than identifying what temporal orientation should be given to action to be qualified as responsible, I follow how "good" relations across time are defined at specific sites. Drawing on Didier Fassin's distinction between "a moral discourse" and "a critical analysis of a moral topic" (Fassin 2008, 339), I do not propose "a code of good conduct" but instead examine "the evaluative principles and practices operating in the social world, the debates they arouse, the processes through which they become implemented" (Fassin 2008, 335). I contend that no temporal orientation given to action is inherently ethical—the ethical nature of action should be an object of empirical scrutiny (Fassin 2008; Ballestero 2015, 2019; Smith and High 2017). The specific temporal orientation given to action is only an effect of arrangements formulated to produce it: therefore, of resolutions of problems of rendez-vous.

In this book, I use the case of nuclear waste to historically and ethnographically examine: What did it entail to undertake action in the present? How were temporal gaps, tensions, and blind spots identified? Through the formulation of which arrangements were relations with futures established? What tests did those undergo, and what trajectories did they follow? At which sites were their terms negotiated and qualified as "good" and "desirable," and by whom? And, perhaps most crucially, what kinds of relations between temporally and spatially defined collectives were put into place, and how, exactly, did these come undone? In following these questions suggested by centering problems of rendez-vous, I build on and reconfigure the theoretical and philosophical debates that lie at the heart of not only nuclear waste policy but also many other challenges qualified as "long-term" by actors in the field awaiting action.

In what follows, I explain how this book builds upon and responds to the limitations of three different strains of thinking that have emerged from and shaped societies' efforts to address challenges facing humans and the planet. Conceptualizations of the Anthropocene encourage humans and institutions to consider long-term futures, yet deep time thinking does not in and of itself equate to taking responsible action. Generational thinking, meanwhile, emphasizes linkages and divisions between broadly defined cohorts marked by the experiences and events that took place during their lifetime. However, its capacities to capture the diversity and dynamism of temporally and spatially defined collectives, along with their complex relationalities, are limited. Finally, thinking about containment often focuses our attention on physical containers. Yet, whether this involves linking the times of life, society, science, justice, materials, politics, the economy, or geology, containment technologies that hold together political and moral relations across times are not limited to physical ones. Furthermore, the "success" or "failure" of such might not rest in their ability to contain without leakages, but rather on their ability to both resist and account for interventions. Ultimately, rethinking the resolutions of problems of rendez-vous through *and beyond* frameworks of deep time, generations, and containment generates new possibilities for approaching nuclear waste futures and, indeed, the future itself.

From Deep Time to Temporal Relations

I consider contemporary societies' attempts to reconcile conflicting and competing temporal depths as a response to the situations with which they are confronted and need to deal. Concerns about the past, present, and future of the planet have been crystallized in the contested notion of the Anthropocene, which refers to a proposed geological epoch marked by extraordinary human intervention upon the Earth. Trying to understand what it takes for actors in the field to tackle a material that remains radioactive for a million years, for which human responsibility has long been recognized, invites reflection on this intervention.

"The origin stories of the Anthropocene" include, among others, the advent of the Atomic Age, the invention of agriculture, the Columbian "exchange" that was the result of the encounter between the New and the Old World, and the Industrial Revolution (Yusoff 2018). Whether or not scientists may reach a general consensus about the powers that move geological forces, and whether the causality between nuclear weapon testing and the shift to a new geological epoch can be demonstrated, humanity is undeniably in the midst of a turning point. To grasp this moment, social scientists from multiple disciplines are engaged in a moral enterprise to equip human beings—ordinary citizens, experts, or institutional actors—with the capacity to think beyond what is temporally and spatially situated proximate to us, whether this involves "the planetary" or "deep time" (Saraç-Lesavre 2015; Yusoff 2018; Chakrabarty 2018, 2021; Clark and Szerszynski 2020).

Among those criticizing "presentism" and the lack of temporal depth in the social sciences and advocating for more "deep time" thinking is anthropologist Richard Irvine. Irvine suggests that concentrating on the present can only produce "a temporal disjuncture from the expansive frame of natural history" (Irvine 2014, 160). To connect "the time horizons of lived activity and the scale of these natural processes" (Irvine 2014, 161), Irvine calls on social scientists to orient their gaze toward their empirical fields of study, not only to repopulate them with more-than-humans but also to introduce relevant temporal depths into them (Irvine 2020).

In *Deep Time Reckoning*, anthropologist Vincent Ialenti responds to this call (Ialenti 2020). Condemning the conditions that "trap billions of minds in the narrowness of the now"—which he refers to as the "shallow time discipline"—Ialenti studies a group of regulatory scientists tasked with demonstrating the Finnish national geological repository program's compliance with relevant rules and regulations (Ialenti 2020). Showing how these experts think about deep time and make predictions about geological, hydrological, and ecological conditions tens or hundreds of thousands of years in the future, Ialenti argues that those concerned about the future of the planet need to find novel ways of learning from "expertise-driven long-termism during the Anthropocene" (Ialenti 2020).

I argue that identifying "deep time thinking" as an endpoint in a certain way dismisses the relational nature of time(s)[8] and accentuates the very temporal gaps it aims to overcome. What it means to account for "deep time" for the resident of a small town who is concerned about ensuring that they and their descendants have a "good" life in that specific place on Earth is not necessarily the same as what it means to account for deep time for regulatory scientists, who are concerned about the capacity of that place's underground to act as a durable infrastructure today and for the next million years. Unless those different conceptions of deep time are recognized and connected, the present cannot be attached to the future. I do not consider one way to construe deep time as morally and politically superior to the other. Rather, analyzing specific problems of rendez-vous generated by the efforts to deal with US nuclear waste, represented here as a nexus of challenges that stretch *varyingly* far into the future, draws attention to the actors and actions that can link or delink different time scales and their inhabitants.

"Intergenerational" Responsibilities and Values

The notion of "intergenerational responsibility" has been central to the US nuclear waste policy. In the late 1970s, the Carter administration formulated its policy around a single overarching principle: "the waste generated by nuclear power must be managed so as to protect current and future generations" (Interagency Review Group 1978). In the disciplines of sociology and anthropology, the concept of "generation" has mainly been formulated as a quest to analyze its substance. Foundational to the study of generations is sociologist Karl Mannheim's theory of "The Problem of Generations" (Mannheim 1952). To understand how social time accounts for change in the social fabric from one period to another, Mannheim argued that the problem of generation was a problem of consciousness: how people who come of age in specific historical periods understand their experiences, and how knowledge is transmitted, received, and renewed by the interaction between "new participants in the cultural process" (Mannheim 1952, 292) and the society into which these participants are born and develop—and eventually transform. This resulted in an understanding of generations as distinct entities that both represent and are the result of the transformative events of their time. According to anthropologist Margaret Mead, this distinction sometimes becomes so acute that it results in a "generation gap," which she identifies as happening between the 1940s and 1960s—notably, the formative era of atomic history (Mead 1970).

Beyond the gap identified by Mead, the notion of generation is increasingly deployed to construct new vocabularies of "division" and "collectivism" (White 2013, 235). Thinking with generations entails a certain way of "ordering society, a principle of division by which to make it legible" and—though "an implicit one—it is a choice" (White 2017, 766). It enables the combining of the "scale of the family" and the "scale of the humankind" (White 2017, 769). The former takes hold through a genealogical definition of generation by invoking "our children," while the latter adheres to its philosophical conception by implying a moral obligation toward future populations (White 2017). This double movement links the "past, present and future, and makes the complex sweep of time understandable" (Little and Winch 2017, 137–138), rendering temporal distances governable (White 2017, 765), helping to characterize what distinguishes temporally and spatially defined collectives from each other, and reducing complex and multilayered differences across them by stabilizing them around collectives identified as "generations."

Through the mediation of the concept of "generation," individuals, communities, and institutions not only address ideas about past events but also frame debates about current social problems and, most importantly for my analysis, value temporal relations: What constitutes good and desirable relations between and across "generations," and should relations among generations rely on solidarity or conflict? Should the current generations account for the future of humanity and the planet, as well as their needs and values? Should different generations be regarded as interdependent or sovereign? Should future generations be considered responsible for caring for the undesirable remains of the past? Which "generations" should be considered better off, and based on what criteria? Should there be altruism across generations, and if so, toward which ones? Looking from a distance, once policy legislates acting in the present, societies' relations to futures appear to be resolved (for "good" or "bad"), since the question of whether or not it is worthwhile to act toward the future (and when) has already been answered. Yet, mediating relations between the past, present, and future requires coordinating conflicting and competing times and temporalities, as well as establishing specific temporal relations. As soon as concerned actors engage in efforts to link the past, present, and future in ways that comply with their vision of "good" and "desirable," they engage in practices of valuation. This leads me to examine how actors in the field constantly come to value and revalue the terms of "desirable" temporal relations.

Abandoning the consideration of "value" as a substantive feature, I consider valuation as an empirical act (Dewey 1939; Muniesa 2011, 2014). Valuation, a social practice, involves an analytical engagement with the constitution of values (Helgesson and Muniesa 2013; Antal et al. 2015): that is, as an ongoing, constantly constructed, and contested process (Fourcade 2011; Muniesa 2011; Alexander and Reno 2012; Saraç-Lesavre 2015, 2021a, 2021b). I argue that situated practices of valuation mediate the delineation of responsibilities and the definitions of responsible action, which are produced by—and which produce—specific moral and political reasoning and temporal relations. I contend that defining what constitutes "desirable" and "valuable" relations across time, formulating the "right" course of action, and determining a "good" temporal orientation are all the result of situated practices of valuation.

Crucially, examining situated practices of valuation involves paying attention to many specific details: Which financial relations should be established between those who benefit from nuclear electricity and those who do not? Who should finance a large-scale, long-term nuclear waste program? For how long? For which waste: past, present, and/or futures? Which present and future relations should the disposal of nuclear waste at a deep geological repository establish with the community located on its surface and its deep underground, and for how long? Who does that "community" include and exclude? Which budgetary mechanism should be put in place to enable the completion of such a long-lasting project, while maintaining it as accountable to current political institutions?

The processes of valuation implicated in answering these questions locate the constitution of temporally and spatially defined collectives at the heart of problems identified as "intergenerational" ones—that is, they delineate gradations of difference in terms of responsibilities and rights, resources, and vulnerabilities. Seeing this can illuminate important differences that broad-stroke generational thinking can overlook (e.g., power differentials between groups or the importance of geographic place). Concentrating on the identification of problems of rendez-vous and their tentative resolutions allows reflection on how "intergenerational" concerns are accounted for (or discounted) and, as importantly, translated into action across different sectors and timescales—whether they relate to climate change, care for the vulnerable, public debt, pension schemes, environmental degradation, or, as here, the remains of the nuclear age. Using the vocabulary of *rendez-vous* entails emphasizing how, in any given specific situation, temporalities, responsibilities, and values understood as needing to be held together are neither pregiven nor out there. "Times" to be bridged; "responsibilities" to be delimited, distributed, and assigned; and "values" to be enacted only become explicit once concerned actors and their existing institutional, political, technoscientific, legal, and economic capacities are put to the test. Therefore, the concept of the problem of rendez-vous is an invitation for social scientists to pay specific attention to the ways in which actors in the field continuously try to hold temporalities, responsibilities, and values together with the means at hand.

Containment as Holding Together

Foregrounding the US nuclear waste program's problems of rendez-vous, I draw attention to a broad landscape of containment that stretches far beyond the scientific realms of geology, science, materials, and technology that so often dominate conversations about the challenges of dealing with nuclear waste. Soon after I started this research, I became a member of the OECD's working group on Memory Practices & Knowledge Management in Nuclear Waste Management. During the meetings, national delegates discussed how, within the context of their national nuclear programs, relations between the present and far futures were to be contained in the technologies they designed to warn people in the far futures about the dangers of radioactivity. At one of our meetings, the French delegate proudly displayed the technology the French national program had developed in collaboration with archaeologists, geologists, material scientists, engineers, linguists, and anthropologists. It was a marker made from two thin disks of industrial sapphire. Data was inscribed in platinum onto one side of one of the disks, before the other was placed on top to form an "information sandwich." The disks were then molecularly fused together. The hardware was then immersed in acid to test the durability of the material and to simulate its aging. The material was expected to remain intact for at least a million years.

The Fukushima disaster suddenly destabilized this undertaking, however, as ancient tsunami marker stones failed to warn future generations of rising water levels. Designed over a hundred years ago, these stones dotted the landscape in certain parts of Japan to ensure that future generations did not build below a certain height. Tsunami stones contained materials and information, but more importantly, they contained the resolution of a specific problem of rendez-vous: a sense of moral and political duty felt by one *temporally defined collective* to imagine means to warn others who were both to exist and to act at different moments in time. The language had not changed, and the markers remained materially intact, but contemporary Japanese society had judged that the message contained in the markers[9] was obsolete. Puzzled by this development, national delegates started questioning the utility of their efforts.

What even is containment? What does it mean to contain? History of science and STS-informed scholarship have focused on what it is that containment practices do, often concentrating on their inability to fully contain.

Yet, as demonstrated in Peter Galison and Robb Moss's seminal documentary, *Containment*, containment can be "good enough" (Galison and Moss 2015). Geological repository projects, for instance, do not attempt to achieve perfect containment but rather aim for an "acceptable level" of risk of potential harm from radioactive matter for humans and the environment, achieved at "acceptable" costs. Even if containment technologies are doomed to leak, seep, or overflow, this does not mean that they are useless or not productive.

Sociologist Michel Callon's (1998) reflection on the "logic of framing/overflowing" is helpful for thinking about the relation between containment technologies and their leakages. In his work on the conception of economic markets, Callon suggests that framing is a process that involves a set of concerned actors agreeing "on the frame within which their interactions will take place and on the course of action open to them" (Callon 1998, 249). However, performing reality within these relations does not erase all other relations contained inside them. Overflows allow what had previously been excluded into the framing and reveal relations, concerns, and actors that had initially been set aside. Therefore, while overflows are often described as accidental leaks or anomalies by those who design the framing or those who put it into action, Callon insists on the opposite: "overflows are the rule and framing is a fragile, artificial result based on substantial investments" (1998, 250). Following this invitation to consider framings as costly, always imperfect, and incomplete, and building on recent work that moves beyond enclosure paradigms, I explore what containment(s) do(es) beyond enclosing,[10] and what efforts to contain (such as the Yucca Mountain repository project) "provoke," even when they are qualified as a "failure" (Lezaun et al. 2013; Schoot and Mather 2022).

I consider containment as an experimental method for future-making and adhere to the argument that containment technologies are not passive holders (Lezaun et al. 2013): they actively grasp and rearticulate, and they are always entangled in dynamic processes of world-making (Sofia 2000); in that sense, they should not be understood simply as technologies of enclosure (Schoot and Mather 2022). Schoot and Mather highlight how containment technologies operate "simultaneously" and "in concert" to "hold something together" beyond "helping to hold something inside" (Schoot and Mather 2022, 939). Pushing this insight further, I explore how containment technologies operate in the US nuclear waste program by focusing on the identifications and trajectory of the resolutions of its problems of rendez-vous, while demonstrating that what was to be held together was never solely spatial but was also temporal, moral, and political.

Containment technologies to mediate resolutions of problems of rendez-vous are numerous. They are the results of efforts to make far futures present and calculable, to establish specific debt and credit relations across temporally and spatially defined collectives, to assign responsibilities related to nuclear waste in "good" and "desirable" manners, to estimate the present "value" of future risks and benefits, to develop "sustainable" and "accountable" budgetary mechanisms and management organizations, and, no less importantly, to reassemble geographies, geologies, and communities around the "right" human and more-than-human relations. Therefore, they sometimes take the form of a monetary figure, a pot of money or a rate, and other times the form of a piece of legislation, a contract, a financial report, a regulatory requirement, a concrete canister, a law, a geological formation, or a business plan. All these iterations of containment and more to resolve problems of rendez-vous will be seen in this book.

Overview of the Book

To examine the unique characters that the resolutions of problems of rendez-vous assume in dealing with nuclear remains in the context of recent iterations of the US nuclear waste program, I adopt a multisited methodological approach, navigating among several interconnected sites at which specific arrangements were formulated. This book is based on fieldwork that took shape as I worked continuously to "notice" the "troubled stories" (Tsing 2015)—narratives of temporal relations, capacities, and arrangements under strain—generated by the formulations, trajectories of resolutions of problems of rendez-vous, and the difficulties that concerned actors faced in maintaining what were once construed as "good" and "desirable" relations across time as they attempted to implement nuclear waste policy. I bring these sites into the same frame of study, explain their relations, and show each element's contribution.

I consider the research site as an assemblage of relational ties within a network (Latour 2005). The book unfolds around containment technologies that only became visible with the demise of the Yucca Mountain project: a set of financial figures and instruments, a fund and a congressional budgetary mechanism, and the only operating deep geological repository for nuclear waste in the world. Each of those was once formulated to resolve a specific problem of rendez-vous and to contain responsibilities, values, and temporalities in their own ways. *Nuclear Remains* is therefore thematically arranged into three parts, each consisting of paired chapters that allow the reader to travel through the lifetime of these containment technologies and how they resolved problems of rendez-vous.

Chapter 1 explores the following question: Who pays on a "date with destiny"? It opens with a lawsuit filed by the National Association of Regulatory Utility Commissioners (NARUC), representing private nuclear utilities, against the US Department of Energy (DOE) in 2010. The plaintiffs claimed that, in the absence of an ongoing nuclear waste program, the DOE had no reason to continue to collect the "nuclear waste fee," a unit charge that had been paid by nuclear electricity consumers since 1983. In 2013, the US Court of Appeals ruled against the DOE. This decision suspended the nuclear waste policy's overarching principle of making generations benefiting from nuclear energy financially responsible for their waste.

But how could such an exceptional responsibility be calculated and contained in a unit charge? How were the terms of that peculiar financial responsibility translated into a formal arrangement and distributed among which temporally and spatially defined collectives? Using archival and institutional material from the DOE alongside interviews with relevant actors, I take the reader back to the formulation of what was once described as a morally and politically desirable distribution of financial responsibility. First, I interrogate how the almost indefinite financial responsibility associated with nuclear waste was rendered calculable and then distributed among the collectives that this financial mechanism rendered "responsible." Second, I highlight how the resolution of this specific problem of rendez-vous, synchronizing and coordinating temporally and spatially distanced actors and actions, produced a specific moral and political reasoning.

Chapter 2 brings the reader back to 2008. In its final official economic assessment, the DOE estimated that the completion of the Yucca Mountain project would cost $96 billion and last until 2133. To calculate that number and to estimate whether the fees collected from current nuclear electricity consumers would be sufficient to cover the program's overall cost for over a hundred years, the Department relied on a specific accounting convention: the discount rate. Discounting (or, in other words, determining the time value of money) entails valuing the future through the gaze of a financial investor by rendering costs and revenues with different temporalities commensurable. It is both an extremely contested and yet an almost universally used tool in both policymaking and policy implementation. Depending on the discount rate adopted, discounting acquires the power to decide the fate of federal programs and regulations.

The chapter analyzes how the DOE's role has shifted over time from an explicit moral and political positioning to one that has delegated economic uncertainty to calculations and technical expertise. The chapter is also situated within larger current scientific and political debates surrounding the definition of "the right" discount rate to be used for the evaluation of long-term environmental projects (e.g., climate change mitigation and environmental remediation undertakings). I demonstrate the prominent role that this financial instrument has played in attaching the present to futures and in resolving the problem of rendez-vous generated by the time value of money. The discount rate relies not only on assumptions about future states of the economy and society but also on definitions of who present and future "generations" are and will be.

Building on archival and institutional material, interviews with relevant actors, and a field trip to Nevada, chapter 3 starts in late 2009, when the US Senate Majority Leader Harry Reid, a senator from Nevada and a long-term opponent of the Yucca Mountain project, announced that he had reached an agreement with the White House on the budget to be attached to the Yucca Mountain project. Following this, Secretary of Energy Steven Chu requested no funding for the Yucca Mountain project for the fiscal year of 2011: instead, he only allowed funds for its closure. A project meant to contain radioactivity for a million years was terminated in a single fiscal year: that was how Yucca Mountain ended.

Going back to the creation of the Nuclear Waste Fund in the 1980s shows that this is the opposite outcome from what the framers of the policy had initially intended. Though the Nuclear Waste Fund was quite different from the rock bolts or titanium shields that engineers designed to ensure the containment of nuclear waste, this funding mechanism was nevertheless a technology that was critical to the project's eventual fate. The original design of this program aimed to resolve the problem of rendez-vous of reconciling the contradicting values of providing immunity from short-term political alterations while maintaining accountability to existing political institutions. This was a difficult balance to reach for a project that was planned to last for over a century. Going back to the early debates surrounding its creation, I interrogate how, using existing budgetary instruments, those "conflicting promises" were reconciled through a particular funding mechanism that would have important consequences later.

In chapter 4, I follow the trajectory of the Nuclear Waste Fund. Once translated into action, the technical and political measures required to keep the Fund immune from short-term political alterations slowly evolved: first, as the US nuclear waste program was transformed into a line in the annual US budget (thus competing with other budget items) and, second, by transforming the process of budget appropriations into a site in which the Yucca Mountain project's capacity to be au rendez-vous of contemporary US society's needs and concerns was constantly scrutinized. The chapter explores how each of these processes took place and how budgetary instruments began to reshape not only the organizational time of the project but also the materiality of work onsite. It shows how the resolution of a problem of rendez-vous, once inscribed in the Nuclear Waste Fund, with the aim of keeping the nuclear waste program immune from and accountable to political institutions, was dissolved.

Chapter 5 moves the reader to the city of Carlsbad, New Mexico, where a group of local actors have been fighting to maintain their city's attachment to nuclear waste futures, proposing it as an alternative site to Yucca Mountain. As of 2012, the Waste Isolation Pilot Plant (WIPP), a geological repository in a salt formation that was being used for the disposal of military nuclear waste, appeared to local actors to be approaching the moment when it would stop receiving new waste. Unlike Carlsbad's subterranean landscape, which will remain radioactive almost indefinitely, local actors feared that with the potential conclusion of WIPP's mission, the city's attachment to the nuclear futures would come to an end.

The previous resolution of the problem of rendez-vous, in which the military's nuclear waste was physically contained at Carlsbad's underground, has generated another problem of rendez-vous: how to avoid the transformation of the community living on the surface into the remains of the nuclear age. As long as Carlsbad remained attached to nuclear futures, local actors expected to maintain a certain authority over their destiny and their territory, rather than surrendering them to radioactive matter or the vagaries of deindustrialization. In this chapter, I describe how this group attempted to resolve this problem of rendez-vous by engaging with the long and laborious political, technoscientific, and regulatory processes to physically isolate more nuclear waste in the depths of the Salado Formation. In this resolution, underground, salt, and science were to play a particularly important role.

In chapter 6, I first turn to the way in which these local actors staked a claim to their community's "unique value" for the nation by reaffirming, reappraising, and reinvigorating Carlsbad as a distinctive political assemblage. They did so, in part, by denouncing the so-called "Not in My Backyard" communities that oppose the siting of nuclear projects in their vicinity. I contend that local ambitions to act in "enterprise form" as "free and autonomous agents" reveal how the political and moral reasoning that had once guided the formulation of the national nuclear waste policy has been transformed, as notions of responsibility are transferred from the federal government to private institutions and individuals. In this evolving moral regime, individuals and communities, detached from their dependence on the federal government, acted "responsibly": not only to enhance their own welfare but also to enhance the welfare of the nation as they saw it.

Then, I unpack how the political, technoscientific, and temporal challenges local actors faced in bringing more waste at WIPP's underground encouraged some of them to think of an alternative way of resolving Carlsbad's problem of rendez-vous. This alternative involved the containerization of nuclear remains that would hold temporalities, responsibilities, and values in a completely different manner, a development that raises questions about how the country's problem of rendez-vous with the physical isolation of nuclear waste should be resolved: should it be resolved via creating what opponents describe as a "parking lot" for nuclear waste storage, or a deep geological repository in the depths of the Earth?

I Valuing Far Futures in the Time of Nuclear Economy

1 The Nuclear Waste Fee: Containing Indefinite Responsibilities?

The US Department of Energy's most recent update of its Analysis of the Total System Life Cycle Cost of the Civilian Radioactive Waste Management Program was released on August 5, 2008. It estimated that the Yucca Mountain geological repository project, as then planned, would have cost $96.2 billion (in 2007 dollars) from its beginning to the closure of the repository. The full cost of the program was to be covered by the generators of nuclear waste (US Department of Energy 2008). Accordingly, private nuclear utilities charged their consumers a unit charge to cover the full cost of the disposal of civilian-spent nuclear fuel. These funds were deposited in the government-managed Nuclear Waste Fund. By 2010, ratepayers (users of nuclear electricity) had paid the unit charge for nearly three decades (about $750 million a year). With investment income, the Fund's balance exceeded $24 billion.

Following the abandonment of the project, numerous industry organizations, mainly representing private nuclear utilities, filed lawsuits against the Department of Energy (DOE) in federal courts.[1] The plaintiffs asked the agency to justify, in the absence of an ongoing repository project, why nuclear utilities were continuing to pay the unit charge and what it was they were paying for. Examining the financial assessments historically produced by the Department of Energy (fee adequacy and total system life cycle cost) over the lifetime of the US nuclear waste program, I noticed a striking pattern: the amount of the unit charge has remained the same over the thirty years that it was collected, including the period prior to the designation of any specific community, geography, and geology to host a geological repository project. For me, this raised an apparently very simple question: What was this unit charge to contain? This pattern led me to examine financial containment as a distinct technology for establishing temporal relations, one that operates alongside but differently from physical containment.

The first part of the book is about institutional efforts to contain the exceptional financial risks and responsibilities related to nuclear waste within this unit charge. If the physical containment of nuclear waste constitutes one form of technology of containment, there are many others. Challenges associated with the conception of "good" and "desirable" relations across time became further visible and explicit—and, therefore, potentially vulnerable to redefinition—as institutional actors attempted to translate a given temporal orientation into action. On the one hand, the Department of Energy was tasked with resolving the problem of rendez-vous generated by its obligation to handle nuclear waste by designing a repository that would remove nuclear waste from human time and place. On the other hand, it had to resolve the problem of rendez-vous generated by the organizational time of this massive infrastructure. This involved, among other things, defining who would carry the financial responsibility for nuclear waste, what the amount of that financial responsibility was, how it was to be assigned, and among which temporally and spatially defined collectives.

As economic sociologists and anthropologists have noted, economic arrangements always contain conflicting and competing values, expectations, and temporal relations (Guyer 2007; Vatin 2007; Vatin 2013; Antal et al. 2015; Ballestero 2015; Beckert 2016; Deringer 2018; Dianderas 2024). Economic devices—formulas, scenarios, models, and conventions—actively work to regulate present–future relations, produce valuations of futures, and render distant futures present and actionable (Muniesa et al. 2017; Beckert and Bronk 2018; Esposito 2018; Spears 2019; Doganova 2024; Pilmis forthcoming). In this chapter, I approach the mechanism designed to finance the US nuclear waste program as a technology of containment based on organizing a system of responsibilities and crafting a justified figure formulated to hold parties together.

Following the negotiations leading to the formalization of this monetary figure, I ask: How did the institutions in charge grapple with the challenges posed by formulating the "right" financing mechanism to contain the financial responsibilities associated with nuclear waste? Based on which temporal, economic, moral, and political relations—and, as a result, which negotiations—was agreement on such a unit charge reached in the first place? To unpack the values, expectations, responsibilities, and temporal relations around which this financing mechanism was once formulated, I go back to the period before the formalization of the unit charge (1977–1983) and trace the initial efforts of the US Department of Energy to formalize a financing mechanism that would contain the financial responsibilities associated with the nuclear waste program in a desirable and valuable manner.

To unpack the peculiar relations these economic arrangements established, I join scholars who have demonstrated that the development of calculative devices is not only about performing quantitative operations (Callon and Muniesa 2005; Vatin 2008; Trompette 2013; Ballestero 2015); they formulate embedded systems of responsibility and assign positions within these systems (Muniesa 2000). When I use the term "formulation," I do not mean the development of mathematical formulas, but rather how successive calculations mediate the reconfiguration of political and moral considerations (Miller and O'Leary 2007; Callon 2009; Yon and Mallard 2012). Rather than dissecting equations, I address the technicalities of calculation only insofar as they shape what gets considered a morally and politically "desirable" financial arrangement. Following sociologist Marion Fourcade, I treat monetary figures as equally important to the calculative devices that produce them (Fourcade 2011), since they both operate as precise technologies of containment.

In this chapter (as in this book), when I use a phrase like "the DOE calculated," I collapse all the intricacies of this complex organization into one clear-cut single "agency." This is a purposeful strategy. My objective here is neither to unpack the sociological problems of organizational agency nor to describe the laborious work of the people who intervened in the crafting of the payment mechanism, which would have required another approach. Rather, I approach the financing mechanism as a "text" and attempt to understand the reasoning of its "author" (the "federal agency" that utters it), in the sense of the narrative function, not in the sense of the nuts and bolts of the actual amanuensis. The DOE is a "federal agency," an actor that never remains still: It is constantly being redefined in its efforts to translate the policy principles into action. In that sense, I am not seeking to offer a comprehensive history but rather follow the tests that these complex arrangements underwent throughout their formulations and careers.

The chapter is organized as follows. I start by examining the kinds of efforts involved in physically containing nuclear waste from humans and the environment for deep time, which is a crucial step in making financial responsibilities calculable and, of course, containable. Then, I trace the growing explicitness of the problem of rendez-vous and the provisional resolutions posed by policymakers and experts prior to a formal resolution enacted in national legislation, marking the formalization of what was once considered a morally and politically desirable financing mechanism that would assign responsibilities associated with nuclear waste in a "fair, accurate, and efficient" manner (US Congressional Budget Office 1982). Lastly, I unpack how that resolution relied, on the one hand, on a *system of rendez-vous*, and, on the other hand, on a *nuclear calendar*, two distinct ways of containing ethical responsibilities. Frictions between these systems show the difficulty that experts and policymakers encountered in marking on the calendar when the action should be considered "responsible": at a single moment in the present or over time on a rolling basis.

Making Nuclear Waste Physically Containable

Before unpacking the resolutions of the problem of rendez-vous revolving around the formulation of this financing mechanism, it is important to spell out what it meant to make nuclear waste containable from humans and the environment for deep time, as this had serious consequences for shaping policymakers' decisions about making financial responsibilities calculable and containable. The federal government initially planned to isolate radioactive matter for ten thousand years (the duration required for its radiation levels to decay to that of natural uranium ore) deep inside human-engineered geological repositories.

In fact, deep geological repositories became the preferred option to handle nuclear waste only after several other options were assessed. Outer space, deep sea, or underground (sub-seabed, below ice sheets, or in holes thousands of meters deep down in the Earth) were construed as potential environments to make the by-products of a technological choice inaccessible to humans and the environment. Among these options, outer space disposal was assessed by the National Aeronautics and Space Administration (NASA). One of the concepts that NASA studied involved forming waste into a ceramic–metal material, packaged in special flight containers, for insertion into a solar orbit, where it would be expected to remain for at least one million years (Laporte et al. 1976). This was to be facilitated by a space shuttle that would carry the waste package to a low-Earth orbit, where a transfer vehicle would separate from the shuttle and place the waste package into an Earth-escape trajectory. The transfer vehicle would return to the shuttle, while the remaining rocket stage inserts the waste into a solar orbit. Space disposal was considered the most efficient method to remove nuclear waste from human access for at least one million years. If the NASA space shuttle successfully held its "deep space rendezvous" and no launch pad accidents happened, then space disposal could be the most efficient method to remove nuclear waste from human access. However, the likelihood of each potentially catastrophic scenario was yet to be determined. Ultimately, the space option was discarded due to the large number of flights it required, which was considered to make this option "impractical" and "economically unfeasible" (Laporte et al. 1976).

A geological repository is an ensemble of arrangements that would collectively transfer responsibilities related to nuclear waste from human time and space to geological time and space. Such a transfer involves identifying long-term risks and uncertainties and making them knowable and actionable: How well would volcanoes behave? How would waste packages corrode? Which kinds of thermal and hydrological relations would waste, waste packages, rocks, and aquifers establish? Which kinds of catastrophic events should be accounted for, and how would they impact humans and the environment? What climatic and hydrological conditions would the wastes encounter in a hundred, ten thousand, or a million years? The deep timescales involved encouraged scientists to push the limits of their existing expertise and knowledge to produce predictions.

Three modes of prediction were recurrently used by scientists to produce knowledge about geological repositories' capacity to contain nuclear waste: models, scenarios, and analogues. Modeling involves several steps: formulating a hypothesis based on available empirical data and existing explanations, deriving predictions, and testing the hypothesis through the collection of new data. When it comes to the timescales related to nuclear waste management, testing the hypothesis is not always an option. Scientists often rely on historical observations and geological records that they couple with theoretical conceptualizations to produce predictions (Sarewitz et al. 2000; Oreskes 2000; Macfarlane and Ewing 2006). To that purpose, one of the elements they consider is the future state of the Earth's climate. This involves, among other things, making predictions about future catastrophic floods. For a given specific site, especially one selected for nuclear waste disposal, catastrophic floods do not occur frequently enough to be captured in the past century or so of human observations. This leads scientists to rely on the geological record, documenting how the climate has changed over millions of years. Sediment deposits that are the result of groundwater in caves tell scientists about temperature and rainfall. The remains of flora and fauna in sedimentary layers provide indications about moisture and climate. Flood and lake deposits contain information about water levels and shorelines. Through relying on information about the distant past and the geological record, scientists have extended the timescale of observations. Then, in their attempts to make the far future predictable, scientists constantly resolved problems of rendez-vous arising from the gap in temporal depths generated by historical observation and the geological record.

Scenario-based prediction relies on describing the hypothetical statistical sequence of events in detail and elaborating their potential consequences (Galison and Kruse 2012). This method is particularly used when uncertainty does not lend itself to probabilistic risk analysis. To demonstrate this mode of prediction, let me describe two specific scenarios that have been used to predict geological repositories' capacity to isolate nuclear waste: the impact of a giant meteorite colliding with the Earth and human intrusion into the repository through drilling (US Department of Energy 1980, 5.87). To assess the impact of a giant meteorite hitting the Earth, experts imagined the breach of a repository by a meteorite of 25 m in diameter striking a point on the surface above the center of a repository at a speed of about 20 km/s. Regulatory scientists estimated that its impact would have an energy equivalent of about 3 megatons of TNT, producing a crater roughly 2 km in diameter at the surface and 600 m deep at its deepest point. Choosing this scenario, they relied on information from recorded empirical events: the presence of astroblemes suggested the Earth had been hit by very large extraterrestrial bodies. After defining this scenario, they estimated the impact that such a crash would have on the environment and human health. They assessed the radiation doses that a "reference person" would be exposed to if the repository were breached by a meteorite. This person was a statistically defined one,[2] each time placed in a different region of the United States, close to a different repository site, and living at different moments in the deep future (one thousand, ten thousand, and one hundred thousand years). The scientists then compared this person's risk of death by lightning in the same areas across the same timescales. The conclusion they came to was that "the risk to society remains less than that from lightning and can hardly be considered significant" (US Department of Energy 1980, 5.81). It is through this development of a sequence of imaginable events that risks and uncertainties were valued and rendered containable by the repository.

To assess human intrusion into the geological repository by drilling, scientists did not shy away from imagining the future states of civilization. Following the sealing of the repository, specially designed permanent markers were to be placed on site to warn of the potential hazards associated with drilling operations. One of the scenarios imagined was the potential loss of records through wars or political and social upheavals. Scientists accounted for the complete collapse of existing civilization and pursued the following chain of reasoning:

> The capability for drilling to the depth of the disposal horizon would seem to imply that either the present civilization did not collapse (in which case records of the waste location should not be lost) or that following its collapse, a new technological culture was eventually developed. In the second case, it is reasonable to assume that at least a few thousand years would be necessary to achieve the necessary drilling know-how, so that only the long-lived nuclides and their decay products would be left in the repository at the end of such a period. (Claiborne and Gera 1974)

In both cases, the conclusion was that the repository would be capable of sealing the hazardous by-products of a nuclear society and confining them perpetually in the depths of the Earth. The idea of concealing material from present or future human civilizations was not a novel one. Archaeological evidence inspired those who were tasked with designing permanent markers (Joyce 2020).

Analogues were another mode of prediction that scientists relied upon to identify what constitutes a good and desirable technology of containment. Geologists have long taken advantage of natural regularities that allow a newly studied phenomenon to be compared to a previously known phenomenon. In the 1970s, nuclear scientists considered a uranium mine in Gabon (Oklo mine, West Africa) as a natural analogue to human-engineered geologic repositories for nuclear waste (Cowan 1977; Roth 1977). Noticing a peculiarity in their uranium samples, which contained an isotope found only in spent nuclear fuel produced by human-engineered nuclear reactors, they considered that the Earth's first nuclear fission chain reaction had occurred in this mine (Manaugh 2009). This led scientists from around the world, including from the United States, to travel to the mine to explore how it had contained radioactivity from the biosphere for millions of years, in an effort to identify what produces an "effective" containment technology.[3]

Based on these three modes of prediction, early expert studies commissioned by the Department of Energy determined that a range of from 600 to 1,000 m of earth material should exist between a geological repository and the land surface (US Department of Energy 1980). Such a depth was to provide a barrier between the waste and the biosphere, as well as make it "extremely improbable that wastes in biologically important concentrations would ever reach the human environment" (US Department of Energy 1980, 1.3.4). The isolation of waste in one of several geological formations explored (salt, basalt, granite, and shale) was to rely on a series of barriers preventing the release of radionuclides to the biosphere. These barriers fell under two categories: geological and engineered. Engineered barriers were to withstand the more hostile radiation and thermal conditions encountered initially and ensure total containment of the waste within the waste disposal package during an initial period while most of the fission product with an intermediate half-life decayed. This time period was to last as long as a thousand years, following which the radiation levels and heat generation rates of the total waste were to drop by factors of approximately 1,000 and 100, respectively. Geological barriers were to isolate the waste for at least ten thousand years after the waste was emplaced in a repository. Two important components of the geological barrier considered were the host rock itself and the geological surroundings, which needed to be both tectonically stable and hydrologically noncommunicating (US Department of Energy 1980).

In fact, the very first study on nuclear waste management[4] was conducted by the National Academy of Sciences in 1957 (National Research Council 1957). The study report had concluded that "abandoned salt mines or cavities especially mined to hold waste are, in essence, long-enduring tanks" (National Research Council 1957, 5). Considering that in a geological repository, groundwater is the main mechanism potentially transporting nuclear waste to the accessible environment, two factors made salt formations desirable for a nuclear waste repository. First, because salt is highly soluble in water, the existence of a salt formation suggested the absence of water. Second, should any fractures arise in the salt that might compromise the isolation potential of the formation, it would quickly self-seal, as salt flows would plasticize under the high pressures found at repository depths. Over the several years that followed, specific salt sites were studied for a potential high-level waste repository.

In the end, this was not the kind of formation that was chosen for the deep geological repository project for high-level waste—however, as we will later see in chapter 5, salt formations would remain at the forefront of geological repository options. Some years later, an expert group established by President Jimmy Carter revaluated the role of geological formation in producing a repository's containment capacity (Metlay 2000, 200) and transformed the host rock into one among several elements that would collectively produce containment: the waste, waste packages, rock, aquifers, materials, climate conditions, deformation of the Earth's crust, movement of magma, and human activity.

As passive technologies of containment, repositories were to gradually reach closure, placing nuclear waste out of human presence, engagement, and intervention. As expectations about future states of civilizations, the Earth, and materials were inscribed in the material conditions of the repository project, not only nuclear waste but also responsibilities toward deep time were rendered containable. In other words, the idea that a geological repository would physically isolate nuclear waste from humans and the environment allowed institutional actors to delegate long-term risks and uncertainties to the material components of the repository project and their designers' knowledge, expertise, imagination, and their limits.

Rendering the "Almost Eternal" Financial Responsibility of Nuclear Waste Containable

A growing consensus around the desirability of a geological repository did not imply that the problem of rendez-vous had been resolved but rather that it had changed in nature. Once nuclear waste was construed as physically containable, the Department of Energy started mobilizing expertise to figure out what such a massive project would be made of and how the different physical and financial arrangements that such a project would entail could be punctuated on calendar time. This entailed translating not only expectations about long-term futures revolving around materials and the planet but also the future states of the economy, society, nuclear energy production, and electricity markets into calendar time. To that end, the Department began commissioning expert studies. One of those was conducted by the Office of Waste Isolation, part of Battelle Pacific Northwest Laboratories (PNL).[5] This study report was interesting in that it constituted a first step in the unfolding of the problem of rendez-vous.

To render financial responsibilities calculable, the study began by developing a "reference case." As past studies had resulted in the production of a large body of information about salt as a medium for emplacement, salt was selected as the geological medium for this reference nuclear waste program. The choice of the geological medium impacted many factors, including the time and effort required to mine the rock, the ease in handling the excavations, the pace and density of waste emplacement in the repository, and the "manpower" required to build and operate the repository.

The reference case relied on quite precise expectations about societal futures. The total lifetime of the nuclear waste program was estimated to be forty-five to fifty years, considering spent nuclear fuel would cool down on the reactor site before being delivered to a given repository (Pacific Northwest Laboratory 1978, 3). The initial investments were to take place in the first thirty years and decline toward the end of the operational lifetimes of the repositories. Geological repositories were to be "institutionally and environmentally acceptable," while their construction costs were to "proceed according to whatever schedule is adopted" (Pacific Northwest Laboratory 1978, 43).[6] Existing nuclear power plants were initially designed and licensed to operate for forty years, which was the time span that studies punctuated on calendar time (US Nuclear Regulatory Commission 1978).

The nuclear waste program was to comprise two repositories to ensure that the high initial costs associated with repository construction would be more evenly distributed over time. Inscribing these expectations on a calendar, the study noted that a first repository would be ready to receive nuclear waste by 1988 with a capacity of 45,000 metric tons (MTs), and a second one would start operating by 1997 with a capacity of 100,000 MTs. The US nuclear generating capacity was another important parameter considered; it was to continually grow. The Three Mile Island accident had not happened yet. Remarkably, the study only included a 20 percent contingency of cost, explicitly highlighting that "the use of such a factor" was "not intended as protection against major unforeseen events" (Pacific Northwest Laboratory 1978, 37, 43). The study report made it crystal clear that once completed, it was the geological repositories that were responsible for containing potential long-term environmental and health hazards. At this stage, the resolution of the problem of rendez-vous relied upon the calculation of the total cost of what seemed to be an exceptionally ordinary and on-schedule large-scale infrastructure project.

Distributing Financial Responsibilities among Temporally Defined Collectives

Once the costs were rendered calculable, the next step was to identify who was to finance the US nuclear waste program and how. Formulating the national nuclear waste policy, the Carter administration had given a genealogical and philosophical definition to the notion of "generations." As an indeterminate and abstract group, "present and future generations" were to be protected from the undesirable effects of nuclear power. In 1977, the newly established Department of Energy announced that "those benefiting from nuclear energy should be financially responsible for their waste" (Interagency Review Group 1978). "The present and future generations" were construed as a single group to be protected, while "those who benefited from civilian nuclear energy" were to be financially responsible for the portion of waste their use of nuclear energy had generated. To translate this principle into practice, the DOE announced that it would accept and take title to commercial spent nuclear fuel for "a one-time fee so that the costs of spent fuel disposition could be accurately accounted for in the determination of electricity cost" (Pacific Northwest Laboratory 1978, 1).

But what is a "fee"? It is worth taking a closer look at this fee system.[7] A fee system involves establishing a direct relationship between those who pay for a service and those who provide it. Therefore, calling something a fee versus a tax makes an important difference. Whereas a tax system can be construed as an explicit effort to avoid establishing reciprocity between a specific stream of revenues and a specific service (which may or may not return to the individual or the collective that paid the tax), a fee system involves explicitly attempting to establish that link (Bezes and Siné 2011, 29). It implies charging individuals for the uses they make of government services and aims at generating a similar effect to that of price systems such as a toll system, in which a fee is assessed per passage for a public road.

From the 1970s on, in the United States, the formulation of environmental policies increasingly relied on a reasoning of governance guided by economic efficiency and market incentives (taking, for instance, the forms of "right to pollute," "auto-regulation" by firms, etc.), which increasingly replaced the reliance on regulatory reasoning (taking the forms of norms, interdiction or amends, etc.) (Gunningham 2009). During that period, fees became particularly popular among conservative politicians and those who advocated the privatization of government services (Anderson 1991, 14). In the words of a public choice theorist, Bruce Yandle, who defends governments' increased reliance on the establishment of these kinds of fiscal relations, introducing "economic incentives" fee systems enhances "efficient behaviour" (Yandle 1991, 34–35).

In the case of the nuclear waste program, this meant that through the adoption of a fee system, the nuclear industry could continue to generate nuclear waste for as long as it accepted covering the cost of its management. The cost would then be contained in the contribution required to build and operate repositories, since long-term responsibilities associated with nuclear waste were entrusted to geological repositories' capacity to physically contain risks and uncertainties as soon as they were sealed. In a certain way, this principle provided industrial actors with the "right to pollute" by not aiming at "zero pollution" but instead at an economically efficient means for its economic governance (Fressoz 2013). In that sense, it can be said that the US nuclear waste policy constitutes one of the first economic treatments of an environmental problem with acknowledged long-lasting impacts by any national government.[8]

However, a fee system could take many different forms and rely on different moral, political, and technical reasoning. The potential forms it could take generated debates about what constitutes a desirable and valuable distribution of financial responsibilities. Initial studies commissioned by the Department of Energy suggested the use of a "one-time fee." Based on a one-time fee, each nuclear utility was to pay a lump-sum amount on the metric tons of spent nuclear fuel each of its reactors were expected to generate at the moment of their delivery to a geological repository (Pacific Northwest Laboratory 1978; TRW Energy Systems Planning Division 1978). The nuclear waste program was initially to be financed by taxpayers. Later, as the repositories became operational, the federal government was to be reimbursed by nuclear utilities. The closer the moment of payment was to the moment of delivery of spent nuclear fuel to a given repository, the more certain the cost estimation would have become, and the more fully the US Treasury would be reimbursed. Thus, to establish a desirable temporal arrangement, the emphasis was put on the accurate estimation of the total cost (and that of the resultant one-time fee). Accuracy was expected to increase over time, as repositories were built:

> The one-time fee will not be established until sometime in the future when DOE is ready to enter into firm contractual commitments with utilities, and the actual fee paid by utilities will be that in effect at the time of fuel transfer. At that time, many of the uncertainties will be resolved. (Pacific Northwest Laboratory 1978, 2)

Accuracy was an interesting state for a fee to be in. A sense of progressive adjustment toward a sound figure was expected. These early attempts addressed the task of assigning responsibility as something that could work out unproblematically in the future, yet this did not provide a sufficient resolution to the problem of rendez-vous. Such an encounter was not to take place between abstract and indeterminate "generations" at large but between specific temporally and spatially defined collectives. What counted as an accurate figure had to be the result of a series of temporal meetings between past, present, and future users of nuclear electricity (ratepayers) and taxpayers, and this involved identifying who the temporally and spatially defined collectives that would carry the risk of investment were.

While marking the organizational time of the nuclear waste program on the calendar, the federal government also had to ensure that nuclear utilities would be willing to use the services it provided, as federal nuclear waste services were to be used "on a voluntary basis" (Jacob 1990). This was an intriguing situation. This time, it was the coordination between organizational time and the timing of payment that was at stake, unfolding the problem of rendez-vous in a different manner.

To constitute a national nuclear waste program that would protect "the present and future generations," services offered by the federal government had to be "desirable." This was a direct consequence of the ownership status of commercial spent nuclear fuel. In the United States, civilian nuclear power plants have historically been run by private nuclear utilities, and spent nuclear fuel produced by nuclear power generation remains in their possession. Until 1970, the Atomic Energy Commission had even provided utilities with a "plutonium credit" to use plutonium extracted from commercial spent nuclear fuel in the US weapons complex (US Department of Energy 1999), which operated as one of many subsidies put in place to facilitate the flourishing of the private nuclear industry.

In fact, when commercial nuclear power was launched with the enactment of the Atomic Energy Act of 1954, spent nuclear fuel produced as a result of power generation was to first cool down in spent nuclear fuel pools on reactor sites for at least five years, then be transported to commercially owned reprocessing plants. These plants would extract certain elements from spent nuclear fuel to either be blended with others and transformed into nuclear fuel for civilian nuclear power plants or be used in the weapons complex (Saraç-Lesavre 2021c). In the 1970s, two factors challenged the nuclear industry: the decision of President Carter in 1977 to place a moratorium on reprocessing for economic and political reasons, which eliminated the reprocessing plants that were to serve as temporary storage facilities for private nuclear utilities; and the increased pressure from environmental groups and states demanding the adoption of more stringent safety and security regulations for the operating licenses of nuclear reactors and the storage of spent nuclear fuel (Walker 2009; Stewart and Stewart 2011). The national nuclear waste policy in the making was to meet the concerns of both groups of actors.

While the abandonment of reprocessing resulted in the revaluation of spent nuclear fuel as "waste" in a de facto manner, commercial spent nuclear fuel still remained in the possession of private nuclear utilities. Then, to test the desirability of its services, in December 1977, the Department sent out letters of inquiry to nuclear utilities (US Department of Energy 1978, 7), in an attempt both to evaluate their interest in transferring spent nuclear fuel to federal facilities and to gather data about their preference for handling waste in the "short term."

Here, the definition of "short-term" was crucial. If the utilities chose to expand their on-site storage capacities, rather than transferring their waste to federal facilities as soon as those became available, the short term could extend to almost a hundred years. Nuclear utilities had different technical options for storing their nuclear waste: for instance, they could expand the capacity of their spent nuclear fuel pools or use the "re-racking" method to reconfigure the emplacement of waste inside the existing pools and reduce the space they occupied. If utilities decided to store their waste for a longer time period in the reactor pools, the repositories' operational period would have to be considerably extended, increasing the total cost of the nuclear waste program.

In other words, providing a "desirable" service required utilities to store their waste long enough that they would not need temporary federal storage facilities, but also for a short enough time that they would still use federal disposal services. Indeed, charging a fee only at the moment of delivery would have facilitated reaching economic efficiency. However, this would have entailed a high degree of uncertainty concerning the engagement of utilities with the nuclear waste program. If utilities had not used the provided services, the federal government would have invested in a service that it could not sell and, in return, would have failed to have its taxpayers reimbursed. Formulating the financing mechanism in this way meant that taxpayers would have shouldered the risk of investment.

Early attempts to resolve the problem of rendez-vous were marked by the moral and political reasoning of "incentive-regulation" (Foucault 2010, 354). The timing of payment entailed defining when nuclear utilities would be legally engaged to pay for collective action designed to protect "the present and future generations" from the undesirable consequences of nuclear waste. This relied on a significant effort to accommodate the exigencies of organizational time, which meant reaching a temporal arrangement that not only avoids interrupting the economic activity and protects nuclear utilities from the financial burden of the treatment of their by-products but also functions coherently with the organizational time of a geological repository project.

Making the Values of Temporal Encounters Explicit

The explicitness of the problem of rendez-vous underwent an interesting progression when the one-time fee option was presented to the independent review group responsible for reviewing the nuclear waste policy: the Interagency Review Group (IRG). Created by President Carter and composed of experts from thirteen executive agencies (Walker 2009), the Group expressed various concerns about the Department's focus on a single-fee collection option. They pointed out that the selected financing mechanism required significant taxpayer engagement, as it implied a taxpayer-financed startup for a program for which ratepayers would be financially responsible. It advised the Department to study other options before making any final decisions (Interagency Review Group 1978).

Asked to address this issue further, the Department commissioned a new study by the Pacific Northwest Laboratory (1979) and tasked it with assessing alternative payment moments. The study was to evaluate the role of different temporal encounters in defining what constitutes a desirable way to assign responsibilities at four different moments of payment: at the reactor site during power generation, at the reactor site during fuel discharge, five years before the delivery of spent nuclear fuel to a planned repository, and at the moment of its delivery to a planned repository. The study report evaluated the capacity of each of these four moments to fulfill four criteria.

First, the financing mechanism should minimize uncertain future government expenses, referred to as "outlays." Second, the fee mechanism should minimize the economic impact that the fee collection would have on electricity users. These factors were difficult to reconcile and called for a decision about whether the initial risk of investment should be put on the shoulders of ratepayers or taxpayers. Of course, one should not forget that ratepayers are also taxpayers. Here, the problem was the cost curve of the nuclear waste program. At the beginning of its lifetime, the program was capital intensive. Sharing those initial risks of investment, without burdening either early or late ratepayers or taxpayers, necessitated reaching a delicate temporal balance.

Third, the total cost of spent fuel management was to be fully recovered from those using the power generated by nuclear energy. This was particularly challenging. Depending on the timing of payment, an unexpected cost increase (or decrease) could burden either temporally defined ratepayers or taxpayers. Finally, the federal government was to formulate a financing mechanism that encouraged utilities to use their own interim spent nuclear fuel storage facilities while ensuring that they would use the federal nuclear waste services as soon as they became available. To evaluate which payment moment would better fulfill this expectation, PNL relied on a study that assessed utilities' willingness to pay for on-site storage to avoid delivering their spent nuclear fuel to federal facilities (Pacific Northwest Laboratory 1979, 17), which suggested that early payment mechanisms encouraged the use of federal nuclear waste services.

Ultimately, the study report concluded that none of the options could successfully unite all the expected qualities. It highlighted a dilemma in the way that the different options distributed responsibility: "no fee collection mechanism can guarantee that the appropriate individual customer pays the entire costs for managing spent fuel" (Pacific Northwest Laboratory 1979, 3). This is a remarkable outcome of the problem of rendez-vous. Payment moments situated between power generation and delivery presented trade-offs between the interests of past, present, and future taxpayers and ratepayers:

> Early fee collection mechanisms tend to increase the probability that the individual electricity customer will pay an appropriate rate for the utility to accumulate the fee. . . . Later payment mechanisms increase the likelihood that the utility will pay the appropriate fee in its transaction with the government, but makes it more difficult for the utility to collect its entire costs from the appropriate individual ratepayer. Thus, there is a natural trade-off between early and late payment mechanisms that must be considered in evaluating this criterion. (Pacific Northwest Laboratory 1979, 4)

The study rendered visible an expectation concerning the formulation of a temporal arrangement that would make the users of nuclear energy fully responsible for the cost of the program, distribute that responsibility fairly among those users, ensure that nuclear utilities would use federal disposal services, and avoid burdening taxpayers. Following this new study, it became even more evident that punctuating a specific "optimal date" on a calendar, around which temporalities, responsibilities, and values would unfold in the desirable manner, would not be an easy task to accomplish. While an early fee collection avoided burdening taxpayers and ensured that the federal disposal services would be used as soon as they became available, it relied on a highly uncertain cost estimation. No single temporal arrangement could successfully resolve the problem of rendez-vous for all four criteria.

System of Rendez-Vous Versus Nuclear Calendar

Following the results of the PNL study and the ongoing congressional debates on related legislative proposals, the idea of calculating a "one-time fee" was abandoned. In 1982, Congress enacted national nuclear waste legislation that prescribed the collection of a temporarily defined, per kilowatt-hour fee, starting from the period of power generation. The capacity of the fee to cover the total cost of the nuclear waste program was to be assessed annually, and the fee was to be revised, if necessary, upon the request of the US Department of Energy and the approval of Congress (Nuclear Waste Policy Act of 1982). Rather than waiting for the relevant federal agency to obtain the complete information and to calculate an accurate total cost, the legislation created a relationship of responsibility that could be constantly readjusted. It removed the burden of reaching an optimal temporal encounter and transformed the financing mechanism into a mechanism of temporal compromise: a system of rendez-vous settled the program on an incremental conception of time. The kilowatt-hour fee could be adjusted to future cost modifications and, by being adjusted in this way, could enable fair, accurate, and efficient distribution of financial responsibilities, transforming uncertainty into a desirable state (US Department of Energy 1983a).

Translating the system of rendez-vous into action, the Department of Energy and the utilities were to enter into bilateral contractual relationships and to begin billing the kilowatt-hour fee to their customers (ratepayers) as soon as the contracts were signed. This gave "those benefiting from nuclear energy" a very specific meaning: electricity consumers whose use of nuclear energy was calculable and quantifiable. On the one hand, this enabled nuclear utilities to directly transfer financial responsibility onto their ratepayers. On the other hand, it excluded those who make use of public infrastructures and spaces that rely on electricity generated by the same source, such as local councils, federal agencies, schools, hospitals, roads, supermarkets, grocery shops, or post offices.

However, for nuclear utilities, the problem of rendez-vous was only partially resolved. Starting to charge ratepayers from the moment of power generation meant charging present and future users of electricity for waste that had already been generated. During the congressional hearings on proposed legislation, utility managers submitted several comments seeking to prevent current and future users from being charged for waste generated in the past:

> Since the fee proposed in Title VI of S.1662 would only be collected prospectively, these future payments do not accommodate the existing spent fuel inventories. . . . However, we also believe that a one-time fee should be paid into the waste management fund at the time of delivery of spent fuel or radioactive waste from spent fuel removed from reactor cores prior to enactment of this Act, otherwise, future customers utilizing nuclear generated electricity would pay for disposal of spent fuel from which they have received no benefit. . . . Thus to the maximum extent practical, the actual beneficiaries of nuclear fuel would pay for the disposal costs.[9]

In the formulation of the final contract, the financial responsibility of past users was separated from that of present and future users. The standard contract created a two-category fee: a "one-off fee" and a "generation fee." The first category, the "one-off fee," targeted waste that had been generated before the date of its signature (April 18, 1983). The one-off fee was a sum total of costs calculated per utility based on two criteria:[10] the volume of spent nuclear fuel created during power generation prior to the date of the signature of the contract, combined with the different burn-up levels[11] of spent nuclear fuel. Thus, the effort to reach a desirable encounter also integrated the material differences of reactor technologies and the type of waste each of them generated. The final mechanism separated "already generated" waste from "waste to be generated." In doing so, it sought to avoid putting the outlay of responsibility for waste already generated on current and future users, and it integrated the material differences of different streams of waste into its calculation. Even though the fee was temporarily defined for current electricity users, the financial responsibility of past users was stabilized. Each utility was supposed to pay a lump-sum amount for the waste it had already generated.

A nuclear calendar was thus created: "before the contract" (B.C.) and "after the contract" (A.C.). The kilowatt-hour fee created the opportunity to wipe clean the slate of past responsibility and to produce a certain sense of simultaneity and reciprocity between each kilowatt-hour of electricity consumed by current ratepayers and the spent nuclear fuel it generated. Each kilowatt-hour of electricity consumed was to embed a portion of responsibility. Here, the fee was meant to trace, and be traceable to, a portion of the nuclear waste program. The signing of standard contracts at such an early stage of the nuclear waste program transformed federal waste services into the only option for nuclear utilities. From that moment on, utilities started billing 1 thousandth of USD/kilowatt-hour for each unit of electricity "generated and sold" to their ratepayers through their electricity bills.[12]

The signing of contracts[13] not only bound utilities or obliged them to use federal services but also gave them an important power: that of being a contracting party of the federal government, which made the federal government liable to nuclear utilities for all the policy actions it undertakes with regard to its nuclear waste program. From then on, nuclear utilities were not to pay for waste they generated, but for a service they were to receive from the federal government. Under the existing conditions of regulated electricity markets, this was particularly practical for private nuclear utilities in that Public Utility Commissions enabled them to transfer alterations in cost onto their ratepayers (US Congressional Budget Office 1982). Therefore, the kilowatt-hour fee system relied on the conception of Public Utility Commissions as sustainable institutions that would continue to operate as stable intermediaries.

The contract also punctuated an additional date on the nuclear calendar: a legally binding date by which the first federal repository was to start receiving shipments of nuclear waste: January 1, 1998. Temporal and financial uncertainties related to the nuclear waste program had given rise to a system of rendez-vous. However, this specific date marked on calendar time delimited the scope of the system of rendez-vous to uncertainties associated with construction and operating costs, while treating the delivery schedule as certain. This was in clear contradiction to the moral and political reasoning of the system of rendez-vous, which was specifically designed to integrate time-related uncertainties. The federal government (and therefore the taxpayers) were to be liable should the program fail to start operating on time. The rigid form that the nuclear calendar took led to the further delimitation of responsibilities in time and cemented "the economic and market terms" as the governing principle to "assess government action." It opened the nuclear waste program up to an "economic tribunal" (Foucault 2010, 241–247) and, in a way, resolved the problem of rendez-vous "for the market rather than, because of the market" (Foucault 2010, 121).

Conclusion

Would the system of rendez-vous and the nuclear calendar stand up to the test of time? To observe the test of the unit charge with time, it was not necessary to wait hundreds or thousands of years. In November 2013, the US Court of Appeals ruled in favor of nuclear utilities, which had claimed that the Department of Energy fully breached their mutual contracts by failing to provide nuclear utilities with nuclear waste services in exchange for a fee payment and that there remained no legally legitimate reason for them to continue paying the nuclear waste fee.[14] In the thirty-first year of its collection, in May 2014, the Department of Energy suspended the fee collection (Hiruo 2014). Each time a court rules in favor of the plaintiffs, nuclear utilities are compensated for the program's overrun through the US Treasury's Judgment Fund: therefore, with taxpayer monies (Holt 2021). By 2022, this has run to $8.6 billion (US Government Accountability Office 2022).

When the US nuclear waste policy was first developed, to resolve the problem of rendez-vous generated by the efforts to delimit, assign, and distribute financial responsibilities related to such a large-scale long-term project, political institutions gave a genealogical and philosophical definition to the notion of "generations." These were abstract and indeterminate, pointing to what Jane Guyer has described as "durational ethics of responsibility" (Guyer 2007), and first resolved the problem of rendez-vous by enacting a system of rendez-vous into legislation: a mutual engagement in time toward the "fair, accurate, and efficient" distribution of financial responsibilities. As this started to be translated into a contract between the Department of Energy and private nuclear utilities, this resolution took hold on a punctuated and calendrical time, a nuclear calendar, with "intricately exact, contractual and recorded" engagements (Guyer 2014, 148), delimiting responsibilities in time.

This resonated with what Michel Foucault called the "politics of economization"—the process by which market logics become the primary framework for evaluating government action (Linhardt and Muniesa 2011): building a national nuclear waste program implied designing a desirable federal waste service, incentivizing its use, and distributing the risks of investment among a cohort of past, present, and future nuclear electricity users, while seeking the uninterrupted operation of the nuclear industry and protecting "the current and future generations" from the undesirable consequences of the nuclear age. Should any problem occur during and beyond the licensing period of the repositories, the financial consequences were to be transferred to the federal government, which was given the role of a de facto financial guarantor.

The final form that the nuclear calendar took produced a similar effect to that of "train time" and "coordinated universal time" (Zerubavel 1982; Latour 1997; Galison 2004). It came to dominate other times: the extended network operating the nuclear waste program was to work as a coordinated unity around the nuclear calendar. What had initially been thought of as an ethical principle to make "those benefiting from nuclear power financially responsible for their waste" had transformed into the responsibility of the federal government to fulfill its contractual obligations toward the private nuclear utilities as a "service provider" by a certain date. Whereas the geological repository project was to physically contain nuclear waste from the biosphere for a million years, in formulating the mechanism to finance it, the resolution of the problem of rendez-vous relied on the exclusion of responsibilities associated with such scales of times, along with their transfer to taxpayers at large. Overall, one could argue that the resolution of this problem of rendez-vous was marked by a lack of encounter between two times shaped by radically different ways of reasoning: concern about geological time, on the one hand, pointing to eternity; and the time of nuclear economy, on the other hand, which is incomparably shorter than the former. Here is the current US society "staying with the trouble" (Haraway 2016)—a consequence built into arrangements that paradoxically tried to avoid it. Though physical and financial containment promised to render those on Earth's surface immune from responsibilities related to the remains of the nuclear age, taxpayers were always designated to be au rendez-vous as the ultimate guarantors if and when such arrangements failed.

2 Enacting "Good" Temporal Relations through Discount Rates

The long-term interest rate
Determines any project's fate
At two per cent the case is clear
At three some sneaking doubts appear;
At four, it draws its final breath
While five per cent is certain death[1]

As I type these lines, from a poem attributed to Kenneth Boulding, I watch Queen Elizabeth II giving a speech at the 2021 United Nations Climate Change Conference (COP26), in which she called upon world leaders to undertake imminent action to address the environmental impact of climate change. The queen insisted that "the benefits of such actions will not be there to enjoy for all of us here today, as none of us will live forever." Instead, she explained that climate change action is "not for ourselves, but for our children, and our children's children, and those who will follow in their footsteps."[2] Through this speech, the queen joined the calls to account for future lives, rather than discounting them.

In the United States, the national nuclear waste policy was to account for the far future. A policy that purports to physically isolate radioactive matter from the biosphere for thousands of years assumes those who benefit from nuclear energy have a moral obligation to sacrifice a portion of their well-being to protect present and futures from the dangers of radioactivity. In chapter 1, I deliberately set aside an important element of how that policy is implemented via the fee mechanism: discounting. This chapter turns the focus to discounting, showing how this accounting convention provided a mechanism through which policymakers and experts brought the present and the future into the same financial frame. In doing so, the consequential moral and political calculations that helped set the terms of what "intergenerational responsibility" could look like is articulated in terms of dollars and cents.

Discounting is a key tool that contemporary governments use to define, configure, and adjust relations across time.[3] In a nutshell, discounting involves the process of determining the present value of a payment or a stream of payments that is to be received in the future. Depending on the level of discount rate used, tomorrow a dollar may be worth more or less: if the discount rate applied is zero, a dollar tomorrow equals a dollar today, while with a high discount rate, a dollar tomorrow may be worth nothing today. Therefore, the lower the discount rate, the more money needs to be collected in the present to cover future costs, and vice versa. This technical process has profound implications. Put more simply, discounting operates like a time machine in reverse—instead of traveling to the future, it brings the future back to present-day accounting sheets, reshaping it in the process. With high discount rates, the further in time, the more compressed futures become, until entire millennia of human experience collapse into statistical noise. These calculations, although highly contested, are used almost universally, often to render costs and benefits with different temporalities commensurable and to make the future present and actionable (Beckert 2016). This very fundamental tool of capitalism mediates how moral and political relations across temporally defined collectives are conceived.

A financial instrument that attaches the values of the present to the values of the future, discounting provided nuclear waste policymakers with a tool to resolve what many initially saw as the key problem of rendez-vous posed by nuclear waste policymaking: the question of how much current users of nuclear electricity should be charged to cover the overall cost of this first-of-a-kind, massive program. Embedded in this question were complex negotiations around fairness, accuracy, equity, efficiency, and responsibility, which policymakers attempted to resolve quantitatively through discounting. To resolve the problem of rendez-vous generated by the temporal discrepancy between present revenues and future costs, the Department of Energy (DOE) relied on a simple equation: discounted revenues should match discounted costs. In other words, it brought costs and revenues into the same temporal frame.

Discounting, along with the problems it poses, is not unique to nuclear waste policy. With the rise of climate change mitigation and environmental remediation in global policy agendas, governments have extensively examined the moral, social, economic, political, and technical foundations of discounting when assessing their future investments. Increasingly, a wide range of scientific disciplines question this practice by adopting either descriptive or normative perspectives to unpack the various assumptions it embodies. Some of these critiques attempt to calculate the "right" discount rate, while others demand that the practice of discounting be eliminated. These lively debates between those who advocate for the use of discounting in environmental decisions and those who contest it typically presume a dichotomy between the economy (construed either as "immoral" or "amoral") and morality.[4]

This dichotomy is well illustrated in a historical example that is regularly mobilized by both sides of the debate over the use of discounting. In the late 1950s and early 1960s, the Bureau of Reclamation in the US Department of the Interior used cost-benefit analysis for its dam construction projects. The construction of dams on the Colorado River had created environmental problems, including the flooding of scenic areas, the extinction of several fish populations, and increased water salinity. This analysis, as was conventional, included a discount rate. The Bureau had used a low rate in calculating the costs of the project, contrary to the higher rates recommended in regular government accounting procedures. Commentators seized on this choice.

Defenders of higher discount rates, such as Kip Viscusi,[5] a neoclassical economist, argued that high discount rates are a way to block projects that are likely to produce undesirable and irreversible consequences. Because the application of a high discount rate to projects with long-term costs or benefits tended to result in their cancellation, Viscusi (2007) characterized the practice as "pro-environmentalist." He claimed that adopting lower discount rates prevents future generations from pursuing better solutions and hinders those in the present from using current resources more efficiently. He also argued that when decision-makers choose to engage in capital-intensive, long-term investments with uncertain future benefits, they burden the well-being of both the present and future generations. This is a risk we can avoid, Viscusi and others claimed, by using high discount rates (Viscusi 2007).

In contrast, those who rejected the premise of long-term discounting denounce its tendency to favor investments with short-term costs and benefits.[6] Discounting, in other words, discounted the future. Among many others, Frank Ackerman and Lisa Heinzerling (2002), an environmental economist and a legal philosopher who opposed the use of discounting in environmental policies, argued that discounting is often simply a means for governmental institutions to justify their political decisions. They argued that the use of discounting simply immunized the Bureau from criticism by visibly delegating this political and moral decision to numbers. The Bureau's analysts deployed a low-discount rate to justify their political decision, as they considered that building dams on the Colorado River would be "good" in terms of providing the most water to the most people at the lowest cost for the longest period. The fact that the built dam ultimately produced more large-scale environmental harms than initially assessed, they suggested, was not a consequence of the use of the "wrong" discount rate but rather the result of the "bad" political decision-making and assessment, which discounting helped shield from what might have been stronger public scrutiny.

In this chapter, following American anthropologist Bill Maurer (2014), I am attempting to remain "wary of the juxtaposition of morality and economy itself" (Maurer 2014). Rather than approaching discounting as a binary process, as something that is either good or bad, I instead examine the political and moral values contained in the actual use of discounting over the lifetime of the nuclear waste program, while also tracing the evolution of debates surrounding this very method and the assumptions on which it has been operating. In following Maurer's advice, I join scholars who engage with the intricacies of economic calculations to unearth specific moralities that become inscribed, often without overt intent or public recognition, within the choice of discount rates. When examining the practices of nineteenth-century engineering economists, for example, French sociologist and historian François Vatin (2008) explains that commitments to measurement, calculation, and formalization did not mean that engineering economists discarded political and moral values but rather covered those values in several layers of technical sophistication. The same, I show, pertains to the case of nuclear waste policymaking, in which financing complexities covered processes of making and enacting moral and political values that only become visible through the sustained reconstruction and scrutiny of discounting.

To unpack the plurality of values embedded in the uses of discounting in practice, I trace the career of official fee adequacy assessments produced within the context of the US nuclear waste program, with a focus on how and why these changed over time. After the nuclear waste policy became law in 1982, Congress tasked the DOE with conducting an annual review of the program's financial state. These reviews had two purposes: (1) to assess whether the collected fees were sufficient to cover the estimated total cost of the nuclear waste program and, if necessary, to propose revaluations of the fee by submitting proposals to Congress and (2) to justify the ongoing collection of nuclear waste fees from nuclear utilities. As part of these assessments, the DOE first estimated the total life cycle cost of the program and then assessed whether the collected fees were sufficient to cover that cost—an estimate that incorporated discount rates. These rates were used to bridge the temporal gap between the fees collected from nuclear power generation in the present and the total cost of this colossal, first-of-its-kind project in the future. Therefore, I follow how conceptions of discounting evolved over time, both within the context of the nuclear waste policy and beyond.

As soon as we start tracing the uses of discount rates in practice, we encounter them being put to the test. Figuring out which discount rate to use forced the DOE to render explicit assumptions that otherwise remained unsaid. Sometimes this involved spelling out their political and moral position; at others, it involved deferring those to numbers. In exploring the role of discounting in the nuclear waste program, while following debates surrounding discounting, whether or not such debates penetrated the DOE's assessments, this chapter examines the persons, responsibilities, and relations across time and space contained in the uses of discounting. In the end, it also identifies how the ways that the nuclear waste program's applications of discounting changed over time came to signal the difficulties that current institutions increasingly face in adopting an explicit position concerning their moral and political roles in relation to far futures.

The Nuclear Calendar

These contractual arrangements established a specific temporal framework that would prove crucial for understanding how discounting was used in practice. To understand the role of discounting, it is important to recall the conception of the nuclear calendar. As noted in chapter 1, at the creation of the nuclear waste program in 1982, the building, operation, and completion of the two planned repositories were expected to take around fifty years. By 2033, the containment of nuclear waste would be delegated to geological formations, while the entire life cycle cost of the program would be funded by fees collected by 2014. The cost was expected to be high at the beginning of the program and to decline over time. These assumptions provided a clear nuclear calendar: a program with a beginning, middle, and end. As resolving problems of rendez-vous requires the constant reconfiguration, readjustment, and reinterrogation of assumptions about the past, the present, and the future, as well as the relationship between the three, the nuclear calendar was an important framework for policymakers as they sought more granular financial methods for enacting resolutions.

This nuclear calendar, based on a program life of fifty years, relied on a system of rendez-vous that was to establish an "accurate, efficient, and fair" distribution of financial responsibility across time. And because any assessment of that financial responsibility involved establishing a specific relationship between present revenues and future costs, it included discounting. As soon as the nuclear waste policy was enacted into law and the related contracts were signed, the Department explicitly stated the political and moral position it would adopt vis-à-vis discounting. A discussion of this issue appears in the opening pages of the DOE's very first fee adequacy evaluation of 1983:

> Of specific interest to the reader is the specific discount rate used in discounting cash flows of costs and revenues. A real discount rate of 2 percent was used after examination of several alternatives. The Office of Management and Budget prescribed in its Circular A-94 (March 27 1972) a discount rate of 10 percent to evaluate the measurable costs and benefits of Federal programs and projects extending three or more years beyond their inception dates. (US Department of Energy 1983b)

The DOE chose to use a much smaller discount rate (2 percent) than that recommended by the Office of Management and Budget (OMB) (10 percent). This distinction meant that the DOE's accounts assumed that every $98 collected in the present would be worth approximately $100 at a future date, whereas the OMB's guidelines assumed that $90 would be worth approximately $100 at the same future date. Therefore, the higher the discount rate, the less money needs to be collected in the present.

How did these two organizations arrive at such radically different discount rates? The OMB is part of the executive branch of the federal government. The OMB replaced the Office of the Budget in 1970, in part to prepare and evaluate the president's annual budget. The OMB also had the task of determining how regulatory agencies should assess which projects and programs were worthy of government investment. Such assessments were considered particularly relevant at a time marked by the introduction of environmental and occupational health and safety regulations and the ascent of cost-benefit analysis more generally (Berman 2022). In 1972, the OMB released its Circular A-94 to provide guidance to government agencies tasked with conducting this type of analysis.[7] A first of its kind, the regularly revised circular promoted the use of cost-benefit analysis in regulatory analysis.

Circular A-94's recommendation of a 10 percent discount rate mirrored practices in the private sector. Federal agencies were to assess public projects using "the estimation of the average rate of return on private investment before taxes and after inflation" (Office of Management and Budget 1972). In other words, they were to make assessments based on the opportunity cost of capital. The recommended 10 percent rate relied on an apparently simple reasoning: devoting resources to investments in public projects displaced resources from the private sector. Such a transfer could only be justified if the analysis could demonstrate that the project produced benefits greater than the loss generated by the removal of these resources from the private sector. The government should invest public funds if and only if the gross benefits to the public exceed the opportunity costs in the private sector. The rates recommended by the OMB were initially applied to the cost-benefit analysis for regulations that could be assessed in terms of monetary benefits and whose costs and benefits were market-tradable commodities. From the Reagan administration on, the OMB progressively expanded the logic of cost-benefit analysis to cover nonmonetary benefits, including human lives (Howard and Benfield 1991).

The idea of calculating discount rates on the opportunity cost of capital came from the nineteenth-century German forester Martin Faustmann. Faustmann (1849) suggested that foresters could determine the value of a section of forest land only by evaluating the expected equity value of the forest. He thus coined the concept of "land expectation value," which was to be calculated through a formula that discounts all expected benefits and costs to a common point in time for a periodic management regime. Faustmann's formula, therefore, determines forest land value from the future and from the land's potential alternative uses (Doganova 2018a). This reasoning constituted the foundation for the OMB's recommended discount rates. Based on a similar idea, federal agencies were to maximize the utility of their capital investments. Through its recommended discount rate, then, the OMB not only equated the role of the federal government to that of a private firm but also identified federal agencies as an ensemble of private investors who were competing to access resources.

The DOE rejected this logic. If a federal agency decided to deploy a different discount rate than that recommended by the OMB, it needed to explicitly spell out its reasoning, and this is what the DOE did. The DOE argued that when it comes to the nuclear waste policy, the federal government should hold a different role than that of a private investor:

> That rate was intended to reflect the average rate of return on private investment before taxes and after inflation. That discount rate was not used in this report because a) the disposal of spent fuel and processed waste is a Federal responsibility for which private investment evaluation criteria do not apply; b) today's market and investment conditions are different from those extant 10 years ago; and c) the discount rate noted in the Circular pertains to the evaluation of newly proposed programs, not to those already authorized by the Congress.
>
> In previous reports on nuclear waste fee calculation the DOE used either 6.5 percent or 7.5 percent as the applicable discount rate. These rates, which reflected average interest rates for marketable U.S. Treasury bills, notes and bonds, were used in pricing the Department's uranium enrichment services. Adjusting for inflation, the rates used in pricing such services correspond to the real interest rate of 2 percent employed in this analysis. (US Department of Energy 1983b)

Rather than using the OMB's recommended discount rates, the agency based its discount on the interest rates paid by the US Treasury bills, bonds, and notes.[8] They considered that these investments would provide a low rate of return because they are supposed to carry little to no risk to investors. Individual government investments, after all, are situated within an exceptionally large portfolio. The sheer number of projects being pursued simultaneously by the government acts as a sort of insurance for society against an individual project's failure, abandonment, or cost overruns (Baumol 1968; Arrow and Lind 1970). This choice of rates signaled to investors that federal government was an enterprise that was too big to fail.

At this stage in the nuclear waste program, in resolving the problem of rendez-vous, the DOE adopted a low-discount rate, refusing the OMB rates that encouraged short-termism. In doing so, the DOE explicitly indicated that the nuclear waste policy should be regarded as a responsibility of the federal government, rather than a financial investment decision by a private firm.

Flipping the Calendar

Neither the challenges of this first-of-its-kind project nor its temporality remained static. In 1987, when Yucca Mountain was designated by Congress as the only candidate site to host the country's nuclear waste, its geological formation, "tuff," was tasked to remove nuclear waste from human intervention and isolate radioactivity from the biosphere for thousands of years. The cost of maintaining this facility, once filled and closed, would be minimal—nature, once transformed into an infrastructure, would confine radioactivity almost in perpetuity, to the benefit of those living in the future. In 1995, however, the DOE scientists discovered that water would be moving through the repository system more quickly than they had initially estimated and may corrode the containers.[9] They now warned not only that the oxidized containers would leak but also that radioactivity would follow the water through and out of the repository (Long and Ewing 2004; Macfarlane and Ewing 2006). With this realization, new calculations were set in motion.

This technical performance problem challenged the premise of Yucca Mountain's suitability to isolate radioactive matter from humans and the environment for a million years, putting in danger the resolution of problem of rendez-vous, which relied on transferring nuclear waste from human time and space to geological time and space. Rather than abandoning the project, the DOE's technoscientists decided to counter this problem by designing specific engineering barriers (Ewing 1999). These engineered barriers consisted of enhanced alloy-22 waste packages: a nickel, chromium, tungsten, and molybdenum alloy that could withstand any type of aggressive environment.[10] The engineered system also involved "drip shields" made of titanium to ensure a nonflow environment around each waste package (Metlay 2000; Stahl 2006). The DOE expected that these shields would decelerate any water releases to the rates initially forecasted in the models when Yucca Mountain was designated (Ewing 1999; Long and Ewing 2004; Macfarlane and Ewing 2006).

This change meant that the DOE had to rely on human intervention in the operation of the project. These human-engineered barriers required modifications to the project's safety and security standards, as well as the formulation of new regulations, each of which generated various technoscientific, regulatory, political, and legal debates about the nature of the project and regulatory science (Macfarlane 2003; Shrader-Frechette 2005; Macfarlane and Ewing 2006; Endres 2009). The resulting regulatory reviews, design changes, and legal challenges extended the temporality of the program significantly. By 2001, the closure of the program was now expected to happen in 2119, not 2033 as previously thought.

What interests us here are the implications of this temporal change in terms of *the* nuclear calendar and the discounting practices attached to it. The human-engineered solution was expensive: the DOE scientists estimated that the system's drip shields would require about a third of the world's titanium reserves.[11] The massive range of technoscientific, political, and budgetary problems that the project encountered was increasingly slowing down this one-of-a-kind colossal enterprise and extending its temporality. Engineers were expected to place these titanium shields in the repository during the final closure period. These new technical safety measures also meant that the total cost curve would not decline over time, as initially expected. Instead, there was now a significant cost bump at the end of the project's lifecycle (figure 2.1).

This revision to the nuclear calendar meant that the DOE had to ensure sufficient funds would be saved in the Nuclear Waste Fund to pay for the long-term expenses that would be encountered more than a century in the future (figure 2.1). The lifetime of the program had shifted from a total of fifty years, with the DOE collecting all fees and paying for most of the expenditures in the first thirty years, to one where substantial costs would not arise until 118 years later. This was a big change, although one, it should be noted, that is still dwarfed in comparison by the geological timescale of radioactivity.

The original plan for the repository project included opportunities for the DOE to account for these temporal changes in its assessments. By 2008, the operational period of the Yucca Mountain project alone had expanded to 2133, and the DOE expected the lifetime of the reactors that were currently operating to end by 2047. Therefore, the expected gap between the fee-generation stage from already operating reactors and the end of the

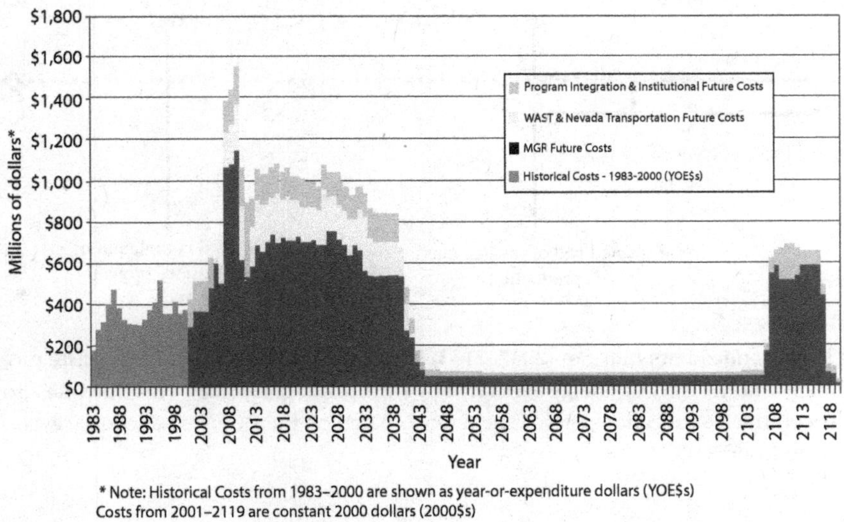

Figure 2.1
Total system life cycle cost profile in 2001 (US Department of Energy 2001).

program's operational phase had expanded from twenty to eighty-six years. This timeline required a temporal adjustment in assigning value to the future. The future of the Yucca Mountain project, along with the safety measures that the project's designers planned would allow those who live up to a million years after the repository's closure, relied on the DOE's capacity to accurately estimate both the future cost of the program and the present value of the kilowatt-hour fees to be collected over time.

The temporal gap generated by the repository's extended temporality and the cost bump at the end of its lifetime challenged the system of rendez-vous that had been established earlier (see figure 2.2). When enacting the fee system into law, policymakers avoided establishing a system that used a predetermined fee to ensure a "fair, accurate, and efficient" distribution of financial responsibility. Rather, fees would be revised iteratively, based on annual assessments. The above dramatic temporal changes modified the relationship between the elements that went into those assessments, including the discount rate.

With the timeline of the project extended, the system of rendez-vous increasingly relied on financial instruments and their capacity to estimate whether the fees collected could cover future costs. In doing so, financial

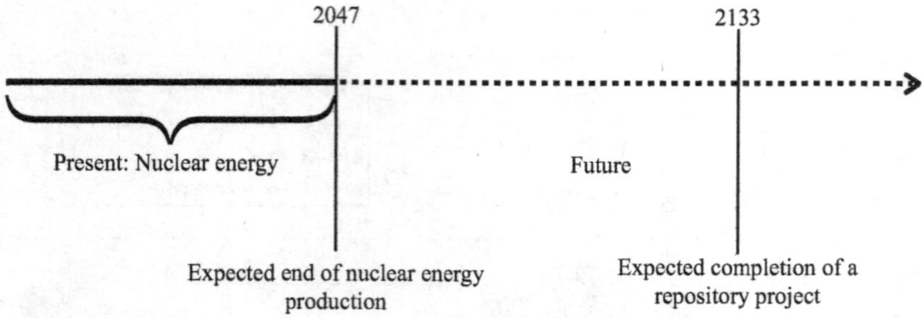

Figure 2.2
The extended temporal gap (2047–2133) between the end of the nuclear waste program and the nuclear energy generated by the operating reactors. This timeline also shows the fee collection period. The figure is based on the DOE's fee adequacy evaluation from 2008.

responsibility for the overall cost of the program had to be accurately assigned to those who benefited from its use: that is, ratepayers (nuclear electricity users). To finance the expected cost bump, the federal government would have to rely on estimates of future costs and revenues, including whether ratepayers were being made to pay enough money to the fund to cover the final expenditures. The discussions concerning discounting not only highlight the DOE and the OMB's divergent perspectives on the treatment of temporal relations but also help us to gradually unearth the temporalities, responsibilities, and values folded into the attempts to undertake action in the present in an effort to protect far futures.

The Person and Discounting

The DOE's initial approach over which discount rate to apply to nuclear waste focused on the nature of the investment: was it a public investor or a private investor? However, the financial calculations associated with the revised nuclear calendar raised an entirely different set of questions about how policymakers in general should solve this problem of rendez-vous and account for the future. Even though the transformation of the nuclear calendar did not alter the DOE's position on what constitutes the "appropriate" discount rate for the nuclear waste program, its revised estimates inspired new research to revisit the principles and assumptions of OMB rates.

To demonstrate how the OMB determined discount rates prevented long-term oriented action, in 2004, two engineer-economists applied those on the changing cost profile of the nuclear waste program and showed how their use would have resulted in the discounting of the costs incurred at the end of its life cycle and, therefore, would have produced the failure of accumulative fees to fully cover the program's total cost (Kenley and Armstead 2004). This questioning did not occur in a vacuum but paralleled the growing interest of economists and policymakers in reconsidering the principles and assumptions on which discounting was originally formulated and had long been used. The standard theory of discounting relied on a specifically imagined person and their expected consumption/investment behavior over time. Increasingly, unpacking that person's behavior became an important object of philosophical and scientific inquiry: Who was that person? What did we know about them? What were their moral responses and principles? What ought those principles be? Responses to these questions were directly linked to the calculation of a given discount rate, which had long remained invisible.

To identify this person, economists started revisiting the work of the founders of discounting. In particular, some turned to the work of the nineteenth-century Scottish explorer and economist John Rae, whom they considered to have established intertemporal choice as a distinct topic of study (Frederick et al. 2002; Rick and Loewenstein 2008). In *The Sociological Theory of Capital*, Rae (1905 [1834]) followed Adam Smith in examining why wealth differed among nations. Rae explained this global discrepancy as the consequence of opposing factors that either promoted or limited "the effective desire of accumulation," including the "bequest motive" and "the propensity to exercise self-restraint" (58). Rae held that these factors differed among countries and determined a society's level of saving and investment. He observed that "when engaged in safe occupations, and living in healthy countries, men are much more apt to be frugal, than in unhealthy, or hazardous occupations, and in climates pernicious to human life. . . . In the West Indies, New Orleans, the East Indies, the expenditure of the inhabitants is profuse" (57). Rae explained that the "uncertainty of human life" and "discomfort [generated by] deferring available imminent gratification" limited savings (120). When the same people came "to reside in the healthy parts of Europe" and avoided "the vortex of extravagant fashion," Rae observed that they were able to live economically (57). He also suggested that "the actual presence of the immediate object of desire in the mind by exciting the attention, seems to rouse all the faculties, as it were to fix their view on it, and leads them to a very lively conception of the enjoyment which it offers to their instant possession" (120). This imagined person preferred early over delayed consumption, was impatient and immortal, and was solely interested in maximizing their own individual satisfaction and therefore discounted the future (which implied their use of high-discount rates).

To put this imagined person to the test, researchers from various disciplines began conducting survey studies and laboratory experiments. To produce knowledge about this "person," some relied on brain scans to visualize their subjects' emotional responses to "reward" (McClure et al. 2007), executed laboratory experiments to test the limits of their subjects' "willpower" (for a review, see Baumeister and Vohs 2003), and measured "delayed gratification" in several psychiatric diagnoses (Wallace 1956; for a review, see Madden and Bickel 2010; Teuscher and Mitchell 2011). Others conducted survey studies to interrogate how their subjects made trade-offs between short-term "tangible" and long-term "intangible" rewards, such as their pension savings or dieting (Polivy and Herman 1992). They characterized their subjects' physiological responses to making such decisions, including their blood glucose levels (Gailliot and Baumeister 2007).

To further challenge the person embodied in discounting, other economists started questioning what determined a person's preferences and intertemporal choices. Arguing that efforts to explain decision-making processes through psychological, behavioral, or physiological factors obscure moral questions, they shifted their interest from what those preferences and choices "are" to what they "ought to be." This included reflecting on a person's relation to others across time and space. Alan Manne (1995), an American economist and operations researcher, for instance, challenged the idea of an impatient, selfish, and immoral consumer. Instead, he conceived of "all humankind as a single agent acting as a trustee on behalf of both the present and future generations." From this premise, he explored the possibility of "an economy in which there is a single agent who acts as producer, consumer, investor, and saver" (Manne 1995, 392). He thus considered all humans spread over time and space as if they constituted a single family. Thus, he argued, when present generations pay for current investments that lead to relative increases in future consumption, they value increments in other people's consumption as if they were increments to their own, and therefore, long-time discount rates had to be low and account for the far future in the present.

American economist Thomas Schelling, who received the Nobel Memorial Prize in Economic Sciences for his work on game theory, pursued similar avenues of thought while partially contesting Manne's conclusions (Schelling

1995). For Schelling, Manne's position meant that "[the] Chinese get nothing from us, but [the] future Chinese we treat as part of the family" (Schelling 1995, 395). Building on the ideas of moral philosopher Derek Parfit, Schelling contested the notion of a hypothetical person who eternally attempts to maximize their own individual utility without regard to morality. Schelling elaborated upon his reasoning by envisioning his own life in the future. He imagined his children growing up, reaching his own age, and then having grandchildren of their own. Then, he pictured his grandchildren also having grandchildren. As a person who "naturally has an inborn preference for immediate consumption," Schelling accepted that he could sacrifice some of his own consumption for his children and grandchildren but considered that it was difficult for him to make the same sacrifice for the grandchildren of his grandchildren. These relations would be strangers to him; even though his genes were "plentiful" in the population at the later date, they were to "spread thinner." Nevertheless, Schelling still considered, "I probably would prefer the benefits to accrue to my own grandchildren rather than to their grandchildren, but I must remind myself that my grandchildren's happiness may depend on their perceived prospects for their own grandchildren, and my 'time preference' becomes attenuated" (Schelling 1995, 396).

Therefore, when present generations undertake action for the well-being of future generations, Schelling argued, they do so not on behalf of their future descendants or because they want to maximize the overall utility of humankind; rather, they do so as a form of altruism (implying the adoption of a low discount rate in assessments). To increase the well-being of a different society, therefore, for him, foregoing consumption made sense if temporal distances were treated in the same way as geographical distances. The question of how discounting was made to resolve the problem of rendez-vous, then, relied on assumptions about, among many other factors, how a given person, situated in time and space, does or should establish their relations with the present and futures. While these moral debates were not reflected in the DOE's fee adequacy assessments, calculations in those very assessments were increasingly filled with efforts to accurately predict and integrate time-related uncertainty.

Translation of Uncertainties into Probabilities

Initially, in the DOE's fee adequacy assessments, discount rates used relied on Treasury bonds that were valued based on a single rate and source. In

1998, however, to account for uncertainty generated by the peculiar and ever-evolving lifetime of the nuclear waste program, the Department started to integrate new parameters. These parameters "used a series of interest and inflation rates, during the period of 1998 through 2042, for investment of income and reinvestment of maturing securities, as opposed to applying a single average rate as in previous analyses" (US Department of Energy 1998).

Furthermore, these parameters began to increase the complexity of the choice of discount rates. Before 2007, for the DOE's fee adequacy assessments, only one economic forecasting company, DRI/McGraw (which later became Global Insight), drove the different estimations of inflation and interest rates. DRI/McGraw determined these rates from values it provided for one- and ten-year Treasury notes and twenty-five-year interest and inflation rates. The more the temporal gap between the revenues and costs increased, the more the assessments relied on an increased number of sources. This diversification was a way for DRI/McGraw to estimate more accurately, despite increased uncertainty, whether the total value of revenues collected from fees would be adequate to cover future costs. As the temporal discrepancy increased, the fee adequacy assessments increased in volume and expanded from just a few pages to more than a hundred in total.

In 2008, the calculations integrated in the fee adequacy assessments were further refined to integrate three additional sources beyond DRI/McGraw for the very first time. This change made long-term economic uncertainties explicit. At this time, the assessments acknowledged the new and uncertain temporal relationship and used new sources to account for the "uncertainty inherent in projecting long range economic conditions" (US Department of Energy 2008).

The DOE's assessments also began to incorporate different economic scenarios against which the adequacy of the fees could be tested (US Department of Energy 2008). Each scenario began with that year's Nuclear Waste Fund balance, added income expected during that year, and subtracted the spending expected during that year to finally reach a year-end balance. This exercise was repeated for each year until the end of the program's lifetime to reach a final balance. In 2008, the DOE tested the future value of collected fees against future costs, year by year, for twenty-eight economic scenarios (four scenarios per source of data; see figure 2.3).

In some of these scenarios, the fee, depending on the discount rate adopted, failed to cover the total cost of the program. The first source came from an econometric forecasting firm that provided three sets of data from an economic growth hypothesis: low, medium, and high. The description of all three hypotheses explicitly excluded the possibility of an "economic crisis," which provided an optimistic view of the future state of the American economy. Other sources included the historical rates of stocks, bills, bonds, and inflation (including the oil shock of the 1970s). The main (and initially only) sources used to estimate the expected value of the thirty-year Treasury bills included the EIA's Annual Energy Outlook, the market yield rates, the OMB's inflation rates, and the short-term interest rate data from the annual presidential budget. Each data set defined a range of rates to apply to the revenues of the nuclear waste fee. The 2008 assessment demonstrated that the fee covered the anticipated future cost under different rates in most scenarios (US Department of Energy 2008). In the following excerpt, the DOE presented its use of scenarios and how these allowed it to consider different futures against which the capacity of the fee to cover the cost was tested:

> The scenario methodology is intended to examine potential outcomes under a broad range of possible circumstances. Changing economic conditions have the greatest effect on financial outcomes and the Global Insight scenarios cover 90 percent of the paths the economy is likely to take, barring an economic crisis. The historical scenario (Ibbotson) reflects data from the oil shocks of the 1970s. (US Department of Energy 2008, 18)

However, the DOE rejected the historical scenario, justifying its dismissal by pointing to the unlikelihood of events such as the oil shocks of the 1970s and considering economic crises as extreme cases. Then, scenarios under which the fund gave a positive balance demonstrated the overall adequacy of the fee (figure 2.3).

Through the deployment of scenarios, uncertainties were rendered "amenable to systematic regularities that can be measured and transformed into probabilities" (Doganova 2018a). Through scenario-thinking and probabilities, the DOE's contractors translated uncertainties into calculable risks and made futures calculable, rendering invisible a whole range of assumptions that enabled the production of those calculable futures.

The longer the nuclear waste program followed its course, the more the DOE's position regarding time-related uncertainty was delegated to technical expertise. In 1983, when the DOE first determined the discount rates, it relied on an explicit, political, and moral claim about the long-term roles

Scenario Group	Economic Scenario	Average Inflation	Average Real Interest	Number of Cost Share Cases with Positive Ending Balances	Number of Cost Share Cases with Negative Ending Balances
Positive	Global Insight Optimistic	1.47%	3.47%	4	0
Base Case (Trend)	Global Insight Trend	1.96%	3.50%	4	0
	DOE Energy Information Administration	2.38%	3.84%	4	0
	Office of Management and Budget	2.30%	3.00%	3	1
	Market Rates	2.47%	2.24%	2	2
Negative	Global Insight Pessimistic	4.00%	2.62%	1	3
	Ibbotson Historical	4.63%	3.03%	0	4
			TOTAL	18	10

Figure 2.3
DOE 2008 Adequacy of the Fee Report (2008).

and responsibilities of those who benefited from nuclear energy. The first fee adequacy evaluation voiced an explicit position when it presented the role of the federal government as that of a public investor. As the temporal discrepancy grew, the DOE increasingly relied on the diversification of sources to demonstrate that it still held the capacity to accurately estimate the fee, while entrusting the treatment of long-term uncertainties to numbers resulting from scenarios, probability assessments, and the private contractors that produce them. Of course, all these economic assessments were taking place within the context of the pre-2008 US economy, a period when a severe economic crisis was not construed as a part of "systemic regularities" to be accounted for.

Risks, Catastrophes, and Relations across Time

This increasing reliance on scenarios and probabilities reflects broader challenges that contemporary governments face in managing long-term uncertainties. Over time, not only has the need to bridge temporal gaps increased, but debates over how governments should approach such challenges have intensified (e.g., climate change mitigation, nuclear waste management, environmental remediation); since the recommended rates were generally high, they were highly criticized for amplifying presentist myopia in policy action.

These critiques gradually began to influence policy. Parallel to the nuclear waste program's evolution, academic and policy discussions revolving around the treatment of the future by discounting proliferated and deepened (see Bazelon and Smetters 2001; Newell and Pizer 2003; Spackman 2004; Zeckhauser 2006; Zeckhauser and Viscusi 2008; Marks 2011). Even though the OMB continued to recommend that federal agencies should use the financial reasoning of a private firm in making investment decisions, in its 2003 Circular, the Office suggested that federal agencies deploy a "sensitivity analysis" to assess the impact of investment decisions on "intergenerational equity" through the application of near-zero rates (Graham 2007). This recommendation marked a small yet important adjustment in the OMB's position: it transitioned from an institution that encouraged the federal government to constantly adopt the role of a short-term financial investor, whose sole goal was to maximize short-term financial utility, to one that recognized, even very modestly, the long-term implications of public investment decisions.

While doing so, the OMB did not use a generic reference to "intergenerational discounting" and instead explicitly referenced the work of Martin Weitzman, a welfare economist at Harvard University. Weitzman's work was situated at the core of the scientific and policy controversies generated by the traditional uses of discounting when it comes to decisions related to intergenerational concerns. Contesting the existing status quo (Weitzman 1998, 2001), along with the use of OMB-recommended high discount rates under time-related uncertainty, Weitzman suggested that economists should account for specific issues at stake, especially when it comes to accounting for long-term catastrophic risks (such as those related to climate change or nuclear waste management).[12]

Weitzman further challenged a long-established assumption about the relationship between time and uncertainty. He argued that the standard theory of discounting was designed to make short-term decisions under conditions of relative certainty, relying on a "thin-tailed distribution" of probabilities under which extreme events, even though catastrophic, have an extremely low probability of occurrence. Rather than discarding discounting, Weitzman suggested deep structural uncertainty and catastrophic risks could be contained in discount rates.

Weitzman underlined that high discount rates (such as those recommended by the OMB) rely on the calculation of the probability of probabilities, and when they are used to make long-term investment decisions, they further accentuate uncertainty. Weitzman suggested that under time-related uncertainty, it is more appropriate to rely on the "fat-tailed" distribution of probability, wherein extreme events acquire a greater probability of occurrence (Weitzman 2009) as the current uses of discounting fail to adequately grasp the risk of catastrophe whose probability of occurrence is below a certain threshold. To escape this dilemma, he argued that "for situations where there do not exist prior limits on damages, [the] CBA [cost-benefit analysis] is likely to be dominated by considerations and concepts related more to catastrophe insurance" (Weitzman 2009, 2). For Weitzman, this entailed the adoption of declining discount rates to better account for extreme risks and the well-being of those living in the future.

Weitzman's approach not only became a source of major debate among economists and policymakers but also revealed the very specific intergenerational relations contained in the choices of discount rates and how they resolved problems of rendez-vous. Among other economists, decision theorist Christian Gollier from Toulouse School of Economics challenged Weitzman's conclusions with a symmetrical model and reached an entirely opposite conclusion: discount rates had to increase over time. Gollier took Weitzman's criteria and used similar reasoning. However, while Weitzman placed the burden of risk on the shoulders of those living in the present, Gollier (2002) put it on the shoulders of those living in the future. Later, Weitzman and Gollier worked together and "resolved" what they called the Weitzman–Gollier Puzzle. By slicing time into temporal sequences, Gollier and Weitzman (2010) distributed risk across time and concluded that, for long-term decisions under uncertainty, declining discount rates are appropriate. Time, long construed as a linear entity, became divisible into parts, and where risk was placed on these parts determined the distribution of roles and responsibilities across temporally identified collectives.

Over time, declining discount rates penetrated the OMB's sensitivity analysis. Furthermore, these rates comprised debates over the choice of discount rates among different national administrations and international organizations.[13] In a manual on cost-benefit analysis, an OECD report by David Pearce, Giles Atkinson, and Susana Mourato (2006) praised Weitzman's work, pointing out that his approach neutralized the "tyrannical" image of long-term discounting (Pearce et al. 2006, 24). In a way, Weitzman's work not only enabled the morally challenged method of discounting to keep its place in long-term policy making but also helped reestablish its authority.

As noted in chapter 1, the system of rendez-vous initially placed financial risks on the present or, more specifically, on those who directly benefited from nuclear energy production. Following the recommendations of the 2003 Circular, the DOE could have included a sensitivity analysis and applied near-zero rates in its assessments. However, the DOE included neither the rates nor the debates about discounting in its assessments.

In 2013, following the termination of the Yucca Mountain project, the DOE delivered its final fee adequacy assessment. In this assessment, the DOE delegated the choice of the rates to an economic consultancy company based in Washington, DC. The company calculated the long-term rates on two bases (see figure 2.4). The first three sources on the table (Global Insight, Ibbotson Historical, and EIA) recorded the average rate for the entire period of each data source. The company then adjusted and projected the average of those rates into the future to obtain specific rates for each of the following fiscal years. For the market yield data, the company used the last date of estimation (2041) and then projected it into 2042 and beyond. Similarly, for the OMB rates, the company took the latest estimation (2012) and projected it into the future.

As seen in figure 2.4, the past, present, and future were each attributed the same status. Whether the rates came from historical data, current rates, or forecasts was irrelevant. The assessment included a brief description on page 157 of Appendix D, which indicated that the economic value of 30-year Treasury bills covered almost 180 years. Stretching the temporal capacities of expertise, these data have rendered invisible the moral and political role these rates have played in linking past, present, and future, and therefore, in resolving problems of rendez-vous.

U.S. Department of Energy Nuclear Waste Fund Interest and Inflation Rate Report

APPENDIX B: LONG-TERM NOMINAL 30-YEAR TREASURY BOND DATA

The long-term Treasury Bond data approximate annual interest rates received on money loaned to the Treasury for 30 years. The Fee Adequacy Assessment assumes that new investments into the NWF will be invested at the 30-year Treasury Bond rate. These new investments are estimated for the entire nuclear waste disposition life cycle and the 30-year rates are provided until 2157.

Table B-1 highlights data provided from the five (5) sources of data presented in this report, as well as the methodology for creating the datasets, and Table B-2 presents the long-term Treasury Bond data sets. The blue highlighted data include data obtained from the initial sources; data in black represent averages based upon forecast or historical data provided.

Table B-1: Data and Methodology for Creating the Datasets for Long-Term 30-year Nominal Treasury Bond Data

Source	Years for Data	Method for Averaging Data and Adjusting to Fiscal Year Terms
Global Insight (includes 3 data sets)	2012-2042	Data from years 2012-2042 were averaged and used as the forecasted rate for years beyond 2042. Then, the data were adjusted to fiscal year terms by adding 25% of previous year's rate to 75% of the current year's rate.
Ibbotson-Historical Market	1971-2011	Data from years 1971-2011 were averaged and used as the forecasted rate for years from 2012 onward. Then, the data were adjusted to fiscal year terms by adding 25% of previous year's rate to 75% of the current year's rate.
EIA	2012-2035	Data from years 2012-2035 were averaged and used as the forecasted rate for years beyond 2035. Then, the data were adjusted to fiscal year terms by adding 25% of previous year's rate to 75% of the current year's rate.
OMB	2012	OMB provided a forecast of 30-year Treasury Bond data for 2012 in their OMB Circular A-94 Appendix C. The 2012 number was used as the future 30-year Treasury Bond rate for years beyond 2012.
Market Yield	2012-2041	Taylor Advisors provided a forecast of 30 year Treasury Bond rate data from 2012-2041. The 2041 number was used for year 2042 and beyond as per the source's guidance.

Figure 2.4
The explanatory table from the DOE's 2013 Fee Adequacy Evaluation.

Across my interviews, most of the former scientists and engineers at the Office of Civilian Radioactive Waste Management (OCRWM) regarded the fee adequacy evaluations and, more particularly, discounting as conventions without consequence and as neutral tools. The task of safely and securely containing nuclear waste depended on technoscientific measures, and the fee was only an intermediary toward their implementation. A retired Yucca Mountain program scientist and manager describes the use of rates as follows:

"They are not safety or security measures. The rates are just numbers to say whether the waste fund is collecting enough funds or not. Are we collecting enough fees or not?"[14]

However, the choice of the rate not only defined the conditions under which the DOE calculated and contained financial responsibilities but also spelled out how current institutions approached long-term risks and uncertainties, construed moral relations across temporally defined collectives, identified persons and delimited their relevant roles and responsibilities, and determined what counts and what does not, based on expectations about the future state of the economy, society, and the Earth along with human behavior.

Conclusion

In this chapter, instead of taking an a priori normative position on discounting and qualifying it as "amoral" or "immoral," I historically and anthropologically traced the uses of specific discount rates throughout the career of the nuclear waste program. Discounting did not solely bring the present and future to the same temporal scale and render the future present, as well as actionable, but also held those different timescales together, though temporarily and partially. To contain financial responsibilities associated with nuclear waste in a kilowatt-hour fee, the US Department of Energy counted not only on the capacity of its planned geological repositories to confine radioactive matter from the humans and the environment for almost eternity but also on the capacity of discount rate to distribute financial responsibilities across temporally defined collectives in "desirable and valuable" manners. Therefore, the adopted discount rates were to resolve a specific problem of rendez-vous: defining the present value of future costs and earnings based on the assumptions about relations among temporally defined collectives, including who they are, under what conditions they live, and how they construe their relations to others.

To discount, one needs to count in monetary terms, and economic and financial reasoning are at play. However, discounting does not inherently contain a selfish, immortal, and short-termist person who conceives of time as linear and the future as prosperous but rather is open to resolve problems of rendez-vous in different ways and, thus, to the formulation of moral and political relations across temporally defined collectives, including those who care equally or more about the future (as well as not at all). This diversity is rendered possible by the flexible and loose nature of discounting, which allows it to contain a multitude of expectations. Through the use of discounting, the future can be construed as catastrophic, prosperous, stable, or stagnant, while human behavior can be conceived as selfish or altruistic toward the present and far futures, and time can be approached as a linear entity or as divisible into parts.

The ways in which discount rates resolve problems of rendez-vous become visible and explicit through the tests they undergo: tests that themselves reflect political ideals and moral values. In terms of the US nuclear waste program, in its early stages, policy experts explicitly called for discount rates to reflect a federal government acting as a long-term public investor. That government, in its desire to govern uncertainty, adopted an explicit moral stance against its conception as a short-term "private investor." Over time, with the increase in uncertainties, the explicit resolution of problems of rendez-vous was replaced by delegation to economic expertise—transferring fundamental questions about long-lasting responsibilities from public officials to private consultants and their seemingly "neutral" numbers. While the Yucca Mountain's capacity to be au rendez-vous of humans and the environment by physically isolating radioactive matter for extended periods of time has been increasingly subjected to public scrutiny, the assessment of the capacity of economic technologies to be au rendez-vous of long-term financial responsibilities related to nuclear waste was delegated to numbers (Porter 1995). This shaped debates and practices of discounting, as well as their relationship to the trajectory of the US nuclear waste program, including escaping public scrutiny.

II Reconciling Project Time and Budgetary Times: The Politics of Budgetary Action

3 The Nuclear Waste Fund: Immune from and Yet Accountable to Political Institutions

I began this project during the termination of the Yucca Mountain project. This termination had been fought for by opponents of the plan in Nevada, which, among others, included activists from the State of Nevada and the Shoshone and Paiute peoples. Some named the bill designating Yucca Mountain as the sole candidate site for high-level nuclear waste disposal the "Screw Nevada Bill"[1] and accused the project of being "pushed down the throat" of the state.[2]

One of those opponents was Judy Treichel. In collaboration with others, Treichel had formed the Nevada Nuclear Waste Task Force just after Yucca Mountain was designated by Congress. Since then, the Nevada Nuclear Waste Task Force has been working as an outreach arm and a contractor for the State of Nevada's Nuclear Waste Project Office, which provides oversight of the project for the State of Nevada. The actions undertaken by this group included, among many things, bringing the State of Nevada's claims of authority over its territory to the federal courts, undertaking legislative action against the project in Congress, and continuously producing counter-expertise that questioned the project's capacity to confine radioactivity away from the biosphere. Yet, despite these decades-long efforts, Treichel says, "We sort of thought, we probably never admitted, but we thought we never would win, we wanted to keep the fight going on, and hoped that the government finally would not want to spend any more money."[3] That is, they ultimately expected the project to be terminated not because of their actions but due to the federal government's reluctance to finance it further.

Treichel's conclusion about her group's protracted fight seemed to be a very surprising expectation concerning such a massive project, long inscribed in the state apparatus. Once designated by Congress, Yucca Mountain's fate was mainly delegated to procedures laid out in the original Nuclear Waste Policy Act (NWPA) of 1982 and in the NWPA Amendment of 1987. First, the Department of Energy (DOE) would conduct environmental impact and technical viability assessments. Second, depending on the results obtained, the Secretary of Energy would recommend a site. Third, the site would be submitted for the president's approval, and finally the DOE would submit the license application to the Nuclear Regulatory Commission (NRC) for a review of its compliance with the NRC regulations and standards.

In fact, after decades of effort, the project was finally abandoned before it had ever reached the stage of hosting nuclear waste, exactly as Treichel had expected, through budgetary action. Budgetary action enabled political action to be taken against a project once designed to remove nuclear waste from the biosphere for thousands of years, while the fate was deliberately delegated to a seemingly separate set of political, technoscientific, legal, and regulatory institutions and procedures.

Taking as a starting point the budgetary action that prompted the abandonment of the Yucca Mountain project, in this part of this book, I argue that understanding the funding arrangements is essential not only because of their ultimate role in bringing this project to an end but also, and more importantly, for understanding the major political and moral role they are set to play in shaping contemporary societies' relationships with the future. Yucca Mountain was a large-scale project of remarkable longevity. When it came to the program's funding, the problem of rendez-vous faced by political institutions and experts took the form of devising a mechanism that keeps the project immune from short-term political interests and shifts while simultaneously keeping it accountable to a democracy's institutions. Policymakers and experts alike strove to formulate a budgetary mechanism that would reconcile these two rather conflicting and competing sets of values.

In each of the next two chapters, therefore, I focus on the formulation and career of the budgetary mechanism put in place to resolve the problem of rendez-vous to bridge the organizational time involved in the construction of this colossal project with the constraints posed by the budgetary times. In this chapter, I return to the first legislative projects, congressional debates, and expert studies. I unpack how the desired balance between these two values was extremely difficult to achieve and fragile to maintain, as well as unpack how the relations between them were constantly revalued, reinterpreted, and redefined as budgetary actions unfolded. In chapter 4, I trace the various tests that the budgetary arrangements underwent over the program's lifetime. I spell out how, through these tests, many of the details specifically designed to maintain the balance between accountability and immunity were transformed to the point of becoming direct political means to end a project originally designed to physically isolate radioactive matter from humans and the environment for millennia. Together, these chapters question: How did policymakers and experts translate and inscribe conflicting and competing values inside a peculiar budgetary mechanism to resolve this emerging form of problem of rendez-vous?

To begin answering this question, I now focus on the subject of this chapter: the Nuclear Waste Fund. The Nuclear Waste Fund (the Fund) is a technology of government specifically designed both to contain the fees collected from ratepayers and to finance the nuclear waste program over the course of its lifetime. This fund is, in other words, a meticulously designed piggy bank from which money would be released by the Department of Energy under the oversight of the US Congress. An analysis of its formulation, therefore, throws into relief how policymakers and experts struggled to reconcile political, moral, and temporal demands associated with the program with the reality of the contemporary political economy, characterized by a growing focus on managing government according to private-sector logics of economic efficiency, effectiveness, profit maximization, and short-term return on investment.

By reconstructing the career of the Nuclear Waste Fund, this chapter offers broader insights into how "political rationalities" are "operationalized" and translated into practice through the mediation of "technologies" and "programs of government" (Rose and Miller 1992). The inscription of political and moral rationalities in a budget mechanism involves "translation" (Latour 1984; Callon 1986), an operation that often creates gaps, discrepancies, and ambiguities (Arnaboldi and Palermo 2011). As Nikolas Rose and Peter Miller note, "Translatability between moralities, epistemologies, and idioms of political power and government of a specific problem space, establishes a mutuality between what is desirable and what can be made possible through the calculated activities of political forces" (Rose and Miller 1992, 182). The initial translations and the ambiguities inherent in them open the path to multiple interpretations, uncertain meanings, and conflicting goals. Furthermore, those who formulate rationalities and those who put them into practice—for instance, budgetary experts and elected officials—often have "heterogeneous and rivalrous" perspectives (Rose and Miller 1992). Examining how specific temporalities, responsibilities, and values have been translated into a budgetary mechanism constitutes an important step in my analysis. Crucially, I also highlight the career these budgetary arrangements have followed in an already operating world, as they constantly enacted, resisted, and were put to the test of ever-changing rationalities, values, and definitions of "good" and "desirable" relations across time.

Negotiations between "Independence" and "Control"

> The waste management program will have an extraordinary longevity. Siting construction and emplacement of wastes will likely occupy a period of 45 years at the most. A repository would need to be maintained in a state such that the wastes could be retrieved for various purposes for an additional period of time, perhaps 50 years, before decommissioning. Thus, the time frame for all these activities is about a century, already an exceptional period of time for most business organizations and longer than the lifetime of most of our government agencies. (US Department of Energy 1984)

While not as intense as debates over the technoscientific, political, and legal procedures to be followed to "permanently" confine radioactive waste from the biosphere, the project's "extraordinary longevity" also generated various debates about the definition of the problem of rendez-vous and the conception of a morally and politically desirable fund that would mediate its resolution.

Expert studies commissioned by the president, Congress, and federal agencies to nurture the debate to formulate a national nuclear waste legislation comprised dedicated sections on its management organization and budgetary mechanism. While the congressional hearings that preceded the enactment of the national policy into law focused more on how revenues would be collected and less on how they would be spent, some policy experts were equally concerned about the resource allocation aspect. Thomas Cotton, director of the Office of Technology Assessment's (OTA's) nuclear waste policy mission, was one of them.[4] Before the House, Cotton expressed the OTA's vision concerning the pending nuclear waste legislation in October 1981, based on the comprehensive report that the OTA was preparing.[5] Cotton tried to lay out what sort of budget the program "should" have. He insisted that the long-term time horizon raised the possibility that typical government funding mechanisms might fall short of the needs of the project.

The Fund was to inhabit two conflicting promises: the "steady and predictable progress towards timely achievement of a long-term goal" and the "oversight and control that is required in a democratic society."[6] He characterized the reconciliation of the two as "one of the most difficult challenges involved in devising a comprehensive waste management program." Creating a desirable budgetary mechanism entailed reaching a balance that would provide the fund with a certain level of independence (by keeping it separate from political alterations so that the program would not be interrupted) and a certain level of control that would keep it under the scrutiny of political institutions (and therefore accountable).

Within this context, an early study specifically commissioned by the OTA and conducted by the Congressional Budget Office's economists interrogated how such a balance could be reached by investigating three traditional congressional budgetary mechanisms: direct appropriations, trust funds, and revolving funds. Among these three, the study qualified trust funds as the most suitable means to reach the desired balance (US Congressional Budget Office 1981).[7] Trust funds were to serve specific purposes and not render funds available for the general purposes of the government, and they could be set up to be accessible without congressional action.[8] These types of funds are understood to generate a relationship between receipts and spending, which differs from that generated by general taxation, which consists of an obligation to contribute to the activities of the government through money or services in a way that excludes any form of return or exchange (Martin et al. 2009, 3).

Whereas taxes generally do not target a specific individual or collective benefit, trust funds explicitly seek that link and are deployed for three purposes: to reduce time-related uncertainty, to ensure funds collected are used for their dedicated purposes, and to avoid the transfer of financial liability to be assigned to a specific group of taxpayers in general. Nevertheless, the federal government's commitment to use such funds for their dedicated purposes is "not subject to an external enforcement mechanism" (Patashnik 2000, 6). An American federal budget expert, Allen Schick, underlines this point: "A trust fund is only as secure as its beneficiaries are powerful" (Schick 2000, 40). In that sense, the use of trust funds per se does not constitute a legally binding engagement sufficient to protect a given fund against political interventions.

Most of the legislative proposals suggested that the revenues from the nuclear waste fee should be collected in a trust fund.[9] There were no major differences in how they proposed it should be spent either. In most of the legislative proposals, spending for the nuclear waste program required congressional approval through the appropriations process; however, this proved contentious. In direct appropriations, spending is authorized through the annual appropriations process and requires the involvement of congressional appropriations committees and, therefore, short-term congressional politics. While the receipts are kept independent from general taxation, spending is attached to Congress and its budgetary authorities. These proposals did not satisfy the OTA, which specifically objected to the usual appropriations process for the Nuclear Waste Fund. Cotton stated that subjecting the Fund to the annual budgeting process was a threat to the availability of funding, hampering the program's stability and continuity both by attaching it to immediate budgetary politics and by impeding the long-term planning that the program required.[10] Giving testimony on those legislative proposals, he pointed out the dilemma inherent in them:

> The predictability of the program depends upon the predictability of the annual expenditures from the waste management fund, not simply the predictability of the revenues going into the fund. By retaining a degree of Congressional control over annual program expenditures, these bills leave open the possibility that, in a period of pressures to balance an increasingly tight Federal budget, some of the expenditures required to assure steady progress on the waste management schedule might be deferred or eliminated—thereby jeopardizing the long-term goal.[11]

These proposals expressed the politicians' difficulty in giving up their authority over both the nuclear waste program and the Fund, which generated a conflict between "a stable commitment to a long-term fixed schedule for a complex technical project such as waste disposal" and "a high degree of external annual budgetary control" mentioned above. While politicians were comfortable entrusting nuclear waste to technoscientists and their capacity to build a technoscientific project to confine nuclear waste from the biosphere for thousands of years, the same politicians had difficulty reducing the control they had over a domain that fell under their jurisdiction: the budget. The nuclear waste program then faced major difficulties in imposing its own temporality in a process that involved the potential intervention of what would be a changing cast of political actors and congressional budgetary authorities.

In fact, while the Nuclear Waste Fund was always referred to as a "trust fund" in my interviews, I was surprised to learn that, in the final legislation, the Nuclear Waste Fund was not formally qualified as a trust fund but as a "separate fund in the Treasury of the United States" that could be "only used for the purposes of radioactive waste disposal activities as defined in the Act." Not a formal classification, the meaning of "separate fund" was left open to interpretation.[12] The original act also included measures to ensure that the program would receive continuous and sufficient funding without interruption. The Fund was subject to congressional appropriations, but unlike the usual direct annual appropriations, it would be subject to a triennial appropriations process. The legislation prescribed that receipts from the fees and spending should be exempt from the Office of Management and Budget's (OMB's) apportionments process—spending in a given year was not to be capped by the revenue of that given year—and included a provision that authorizes the Secretary of Energy, having authority over the program's budget, to borrow funds from the US Treasury to cover any short-term shortfalls in the Fund.[13] With these measures, the budgetary mechanism would maintain a certain level of independence from the temporality of Congress and its short-term politics and be protected from potential short-term funding shortfalls.

Still, by making the program subject to triennial appropriations, the legislation attached the nuclear waste program to the congressional appropriations process and to the need for "coordination" among a diverse range of congressional actors (Wildavsky 1988, 11). It engaged both Houses as they attempted to define the monetary figure to be allocated to each federal program, the budget committees that controlled whether the appropriated funds corresponded to the federal budget targets, the federal agencies that had the budgetary authority to seek funding for their authorized federal program, the Office of Management and Budget that prepared the president's budget, and the president himself. Based on long negotiations between these actors, a budgetary figure for a given program would be enacted in the final budget.

However, the budgetary procedure involved not only congressional actors. As noted in chapters 1 and 2, the collection of the nuclear waste fee was established based on a contractual relationship. A standard contract between the DOE and seventy parties[14] formalized the distribution of responsibility between them. Waste owners and utilities engaged in the payment of a fee (a thousandth of a USD per kilowatt-hour) for each unit of electricity that they "generated and sold," and the federal government engaged in providing nuclear waste services by a specific deadline (January 1998).[15] The revenue collection was subject to strict legal commitments. Therefore, the program was placed not only under the scrutiny of congressional budgetary authorities and institutional oversight mechanisms but also indirectly under the scrutiny of the seventy contractors who each signed a standard contract with the Department of Energy and to whom the Department was bound to provide waste services by a specific date. Should it fail to do so, they could sue the Department of Energy and seek legal relief as soon as they considered that there had been a breach of contract.

During the comments period preceding the formalization of the standard contract, nuclear utilities showed they would be a demanding contracting party. Several industrial commentators asked for the inclusion of a provision in the contract allowing contracting utilities to be attributed "auditing rights" over the managerial and economic efficiency of the nuclear waste program.[16] The Department of Energy responded to this demand by underlining that it "saw no precedent for the private sector to audit the federal government" and added that the DOE's Inspector General, the Government Accountability Office, the Office of Management and Budget, and a public accounting firm would oversee the program. Then, the problem of rendezvous that involved, on the one hand, keeping the Nuclear Waste Fund insulated from short-term politics and, on the other, accountable vis-à-vis political institutions seemed to be resolved, while the reconciliation of those two contradicting and conflicting values relied on a complex arrangement that indirectly included contracting parties.

Developing the management organization that would hold budgetary authority over the program was at least as important for the program as the budgetary mechanisms themselves; to be spent, funding had to be requested. Most of the expert studies recommended that the management organization should take the form of an independent organization or an independent agency (Interagency Review Group 1978; Office of Technology Assessment 1982; US Congressional Budget Office 1982; US Department of Energy 1984). The Nuclear Waste Policy Act of 1982 established not only a budgetary mechanism but also an Office of Civilian Radioactive Waste Management (OCRWM): a single-purpose agency under the authority of the Department of Energy. During the congressional hearings, Thomas Cotton pointed out that the priority during the preparation of the legislation had been given to technoscientific, regulatory, and political procedures, whereas the program's management and funding procedures were considered secondary elements.

He insisted that those points required further reflection and recommended an evaluation following the enactment of the nuclear waste legislation.

In line with that recommendation, Congress prescribed in the legislation that the DOE should undertake a panel study and assess alternatives for managing and financing the nuclear waste program, including the options of the program being run by a private corporation or a federal corporation.[17] As instructed, the DOE organized the study in the form of an independent citizens' panel. Once it completed its study, the panel concluded by highlighting that the siting process and schedule laid out by the NWPA relied on "unusual degrees of detail and prescriptiveness." It stressed as well that an independent agency was the only means to enhance the program's stability, longevity, and continuity, while ensuring its independence from political interference (US Department of Energy 1984).

Declining these recommendations, the DOE responded in a letter emphasizing that the DOE would manage the program by expanding the Office's mission and, in doing so, create a "more stable, more sustainable, and more reactive organizational form" compared to those recommended (US Department of Energy 1985). The DOE was able to reject the recommendations because Congress had only required the existence of the panel and not made its recommendations binding. Facing very strict contractual deadlines, the DOE did not prioritize reaching a balance between accountability and sustainability by hosting these two conflicting values inside a management organization: what counted was keeping the project on time and on budget, which made the resolution of the problem of rendez-vous take a short-termist turn.

Following the refusal of the DOE to account for the panel's conclusions, the OCRWM (or the Office) became the de facto management organization and remained under the authority of the DOE. From then on, it was responsible for all aspects of the waste management program for high-level waste and spent nuclear fuel. Its director, who was appointed by the president and confirmed by the Senate, reported directly to the Secretary of Energy. The program was increasingly placed under the control of the executive branch and was thus subject to the prioritization of its policy choices. In other words, continuous interpretations of the notions of independence and control attached the management organization and the Fund to an administration's political agenda and to what it valued or devalued as a policy priority, while the program's continuity and stability were entrusted to what we will see was a series of politically vulnerable measures.

The Fund's Test with Its Implementer

The vulnerability of the measures put in place to ensure that the program was both immune to and independent from political institutions emerged during the Fund's operationalization. Between the legislation that prescriptively described the Fund and the actual budget requests, the issue of translatability unfolded in a problematic manner. Following the enactment of the law, the Nuclear Waste Fund was brought before one of the major implementers of the US federal budget: the Office of Management and Budget (OMB). The OMB (formerly the Bureau of the Budget) is part of the Executive Office of the President (Schick 2000, 86). It is an institution that prepares the annual budget before submitting it to Congress. It works in consultation with federal agencies and attempts to reconcile federal agencies' budgetary needs with presidential priorities. It is through the OMB's mediation that the federal agencies' budget requests proceed. This mediation does not solely involve defining the amount of funding to be appropriated. As an implementer of the federal budget, the OMB also refines certain definitions, fills in discrepancies, and reinterprets some of the details of the budget in relation to the current president's political priorities. This was also the case for operationalizing the Fund and putting it into action.

As previously noted, the act only loosely defined the Fund as a "separate fund." As part of the president's budget, translating the notion of "separate fund" into practice, the OMB qualified the Fund as a "special fund" and not as a trust fund (US Department of Energy 2001). Special funds are a form of funds earmarked by law for specific purposes but can be subject to congressional appropriation. Hence, despite the prescriptiveness of the act, the Fund was requalified as soon as it encountered its first test with its implementer. This was not all, however. According to the regular audits conducted by the Government Accounting Office (GAO) on the program's progress, the OMB also limited the DOE's budgetary authority over the program expenditures to the Fund's projected annual receipts from nuclear waste fee collection. Thus, as long as this limitation was in effect, the DOE was precluded from using the Fund's borrowing authority to cover the program costs when expenditures in any given fiscal year were projected to exceed revenues.

The objective of the nuclear waste legislation was to maintain the Fund as "self-supporting." With the existing low-level funding available at the beginning of the program, funded by fees that were yet to be received, the Office faced difficulties in financing the program's initial capital investments. As early as 1986, the Comptroller General of the General Accounting Office, in charge of auditing the managing and financing practices of the Office, reported that, in use, the Fund's peculiar properties had been abandoned (US Government Accounting Office, Comptroller General 1986). The Comptroller General further reported,

> OMB's restriction limited the options available to DOE to finance program costs and that based on the existing fee DOE had estimated a revenue shortfall in fiscal year 1987: "Given OMB's restriction on borrowing, DOE noted in its fiscal year 1985 budget request that to cover this revenue shortfall."

Aaron Wildavsky's (1988) *The New Budgetary Process* explores the period during which the Fund was operationalized, marked by the setting of the reduction of the budget deficit as a high priority for the federal government. According to Wildavsky, the role of the OMB changed dramatically over the period during which David Stockman was President Reagan's Director of the Budget. Rather than mainly acting to examine and assemble agency requests in a bottom-up manner and protecting federal agencies against drastic budget cuts during the budgetary process, the OMB evolved into an agency that lobbied the Appropriations Committees for the federal budget, negotiated in terms of spending levels rather than in terms of the content of the programs. Wildavsky pictures this period as follows: "When 'how much' matters more than 'what for,' agencies and Appropriations Subcommittees matter less than the actors (presidents and congressional leaders) who negotiate about totals" (1988, 172). During this period, federal agencies were uncertain of their budgetary authority or whether they would keep their employees, and they were even more reluctant to defend their budget requests in front of the Appropriations Committees so as not to go against the political will of the administration.

In *Deficit and Public Interest*, Joseph White and Aaron Wildavsky describe the "moral economy" that shaped the Reagan administration's public policies. "Reaganomics" presented a view of economics based not on a specific set of theories but rather on heterogeneous theoretical perspectives that enabled President Reagan to "sound like a monetarist, a neo-classicist, or a supply-sider, depending on the subject and his political needs" (White and

Wildavsky 1989, 271). These seemingly contradictory public policies did not find their rationality in their expected outcomes but rather in the way they were associated with a specific set of values: the shrinking of government, the encouragement of profit, and the support of individualism, as well as increased interest in production. To illustrate their argument, White and Wildavsky underline how President Reagan defended the oil-depletion allowance, subsidies for the nuclear industry and synfuels, and auto import quotas. These policy actions were associated with the liberation of nuclear and auto industries from the regulations that had "crippled them." In this way, this political reasoning ascribed a superior moral value to individualism (White and Wildavsky 1989, 273). In effect, the political economic context of Reaganomics shaped the Fund at this pivotal moment of operationalization. In this moral economy, among other things, the budget would be managed so that control over federal spending increased, and the budget deficit decreased. The priorities of the Executive resulted in a process of redefining the Fund, because it eliminated one of the guarantees of the nuclear waste program's continuity and sustainability. Following this new prioritization of objectives, the definition of control took another form. Control meant neither the accountability of the Fund toward political institutions nor congressional oversight but expressed an effort to control the budget deficit.

This conflicted significantly with the moral and political reasoning of the nuclear waste program and the values underlying how policymakers had originally attempted to resolve the problem of rendez-vous. The resolution of the problem of rendez-vous in the initial enabling legislation involved developing a budgetary mechanism that would be sustainable, so that the program's long-term moral and political objectives could be achieved without intervention from short-term politics and accountable so that it complied with the exigencies of contemporary political institutions. However, as soon as the Fund was operationalized and this specific resolution was put to the test in the context of Reaganomics, it proved vulnerable to the very short-termism against which its architects had attempted to protect it.

The Nuclear Waste Fund Becomes Just Another Fund

The nuclear waste program was first revalued through the removal of its particularities inscribed in the script of the Nuclear Waste Fund. Both the DOE's and the OMB's budgetary recommendations rejected the moral premise underlying the resolution of the problem of rendez-vous embedded in the original legislation. However, its encounters with Reaganomics were not over. As the politically vulnerable measures were quickly abandoned, the Fund also became subject to a series of drastic political and legal measures adopted to decrease the federal budget deficit. The first of these measures was the Gramm–Rudman–Hollis Act (GRH), also called the Balanced Budget and Emergency Deficit Control Act of 1985, which defined fixed deficit targets and the sequestration of budgetary resources if the projected deficit was greater than the target (Schick 2000, 292). Wildavsky describes the introduction of these measures under the Reagan administration as follows:

> Ronald Reagan felt that every time he compromised cuts in domestic spending by raising taxes, he did not get in return promised cuts in domestic spending. Though he preferred tax reductions to lower deficits, he did want to balance if (and only if) it could be achieved solely by spending cuts. The GRH formula looked like it might make that possible. (1988, 236)

When the Reagan administration defined its policy priorities and introduced the new budgetary process, it faced the "inertia" of some of the existing funding modalities in the federal budget, namely, entitlements. Entitlements comprised half of the US budget, and they were immune to the new budgetary process. Under Reaganomics, the new budgetary process was introduced based "on the scaffolding of the old" (Wildavsky 1988, 33). This rendered the relationship between the rationalities and the technologies of government more complex and intermingled. Unable to introduce budget restrictions on almost half of the federal budget, the administration had to increase cuts on the items that were subject to direct congressional appropriations, and this included the Nuclear Waste Fund.

However, there was another important detail in the script of the Fund that further facilitated its becoming subject to the short-term politics of budgetary constraints: it was placed "on budget" (Hezir 2011). On-budget funds are accounted for in the annual federal budget, which makes them subject to budgetary measures adopted by the federal government. The 1987 amendments to the GRH placed the appropriations from the Fund under the spending cap applicable to all domestic discretionary programs. Even though the Fund was self-financed and not supported by the US General Treasury, it thus became subject to these constraints. New legislation rapidly transformed the Fund into an ordinary discretionary spending item in the federal budget. Moreover, the Budget Enforcement Act (BEA) of 1990 also imposed new caps on discretionary spending accounts. Integrated into the domestic discretionary appropriation accounts, the Fund was therefore now subject to the spending cap set in the BEA (US Department of Energy 2001). Following this new constraint, the OMB abandoned its practice of setting separate budget planning targets for the Fund, which forced the nuclear waste program to compete against other DOE programs within the same annual DOE budget target for domestic discretionary spending. The OMB's testing of the nuclear waste program transformed it into a rather ordinary federal program. What had once appeared as a separate, dedicated, and single-purpose fund became in practice an ordinary element of the federal budget. Following the difficulties it faced with the OMB, the DOE abandoned the triennial budget requests without ever using them, and the program's annual budget was restricted to the Fund's annual revenues. The Secretary of Energy had to integrate the Office's budget as part of the DOE's annual budget request and submit it for congressional approval.

A program designed to isolate radioactive matter from the biosphere for a million years was increasingly governed by the short-term budgetary priorities of the Executive. The prioritization of the reduction of the budget deficit both devalued the program and eliminated the specific rationale entrenched in the funding arrangements that its architects had developed to shield it from short-termism while maintaining its accountability to the political system. Importantly, it still figured as a separate and dedicated fund. However, with the transformation of its status into a fund like any other, monies collected for the containment of radioactive matter were themselves no longer contained to the extent that the program's architects had envisioned.

Managing Ambivalence: Reconciling Burying Waste "Permanently" with Reducing the Budget Deficit Immediately

The former Yucca Mountain technoscientist, Jane, considered that the Nuclear Waste Fund was being used to cover the federal budget deficit. She describes the situation as follows:

> They see that nuclear waste fund that is sitting there, and they have these millions and when they go to balance the budget, they can say, "Oh, I have all these billions of dollars," and it looks good, and they are gonna let it go like that, and they let DOE pull from it. . . . The only way to effectively run a program is to have

> a set amount of funding [laughs] you know that you can use, and use when you need and not when the Congress lets you have some.[18]

How could a fund conceived to enable the completion of a massive project to contain waste for thousands of years, characterized as "separate" and "dedicated," be used to reduce gaps in the federal budget deficit? The Executive's measures to reduce the federal budget deficit in the short term and the program's demand for high-level spending so that the federal government could maintain its moral and political engagement toward the long-term management of nuclear power remain in conflict with each other. A close look at the program's reports (including the DOE's annual reporting on the nuclear waste program, fee adequacy evaluations, and the Office's reports to Congress) shows that the Fund's transformations involved not only its relation to political institutions but also the "sums of money" recorded in the accounting book.

In the early years of fee collection, the sum of money was limited, and the OMB refused to let the DOE use its borrowing rights from the US Treasury. As the years passed, however, the Fund increasingly showed a surplus balance. According to the provisions of the NWPA, as discussed in chapter 2, the surplus balances in the Fund were invested in Treasury bills to maintain the financial value of the collected funds until they were spent on the program. Political scientist Eric Patashnik highlights that due to their recurrent inaccessibility for the specific purposes for which they were collected, those separate or dedicated funds did not accumulate "real wealth," and when the moment came for the federal government to fulfil its commitments, it often needed to find other financial resources to finance them (Patashnik 2000, 6). As the Fund was placed "on budget," the receipts invested in Treasury bills figured in the annual federal budget. The less the funding was appropriated from the Fund, and the more time passed collecting the fees, the more the Fund's revenues rendered some of the gaps in the federal budget deficit invisible.

Indeed, the Fund did not follow the designers' scenario or remain loyal to its "original intent," but it did accomplish things.[19] A budget was annually appropriated to the nuclear waste program. Translated into Treasury bills, it facilitated the reduction of budget deficit gaps: the invested funds contributed indirectly to financing other federal programs. Moreover, it also continuously figured in all the political and institutional reporting designed to render the financing and management of the nuclear waste program accountable to a number of institutions, and it provided justification for the collection of fees from nuclear utilities and their ratepayers. In this sense, the things done by the Fund were "consequential" (Muniesa 2014).

Reconciling Conflicting Rationalities within the Fund?

The transformation of the budget deficit into a political priority also had consequences for the nuclear waste program itself. From the late 1980s on, the DOE came under increased pressure from nuclear utilities and states that continued to store spent nuclear fuel and high-level waste toward the program's acceleration and the use of the Fund for its specific purpose. The deadline for the contract was approaching (January 1998), and utilities were impatient to deliver waste to Yucca Mountain.

The redefinitions of the Fund were not only observable in monetary figures listed in the accounting reports. Constraints on the Fund also constantly materially reconfigured the Office's mission. According to the GAO's audit report, the low funding process resulted in the slow progress of the scientific investigation at the Yucca Mountain site. As early as 1993, the Office had to reduce the portion of the Fund available for scientific exploration, and the funding received was used mainly for capital investments. By 1993, the GAO was alerting Congress that the Office had used only one-third of its funding for scientific research. The Office was confronted with the need to make choices between operational expenses and capital investments. The GAO signaled the magnitude of the problem as follows:

> If DOE continues to request and allot funds for the investigation project at recent levels, the project could take at least 5 to 13 years longer than planned and increase the total cost of the disposal program. To streamline the project, DOE has compressed the time permitted for various scientific studies and is considering similar measures to reduce costs. These initiatives increase the risk that the investigation of the site will be inadequate. Moreover, they have come at a time when unanticipated technical issues have emerged that could lengthen the investigation project . . . DOE has reduced the scope of the investigation of Yucca Mountain in an effort to adhere to its schedule and is considering further measures to reduce costs. The Department has also developed proposed legislation to change the way the funds are provided for the disposal program so that annual funding would be based on need rather than on ability to compete with other programs for limited appropriations. (US Government Accounting Office 1993)

The low level of funding required the Office to hierarchize its two major activities. This also resulted in tensions between the engineers and scientists working on the Yucca Mountain project concerning "who should receive more funding the following year?"[20] Yet, following the transformation of the Fund in use, the DOE had difficulties increasing the funding level.

To accelerate progress on the project, the Department of Energy began trying to free the Fund from federal budget constraints and the congressional appropriations process. Starting in 1993, the DOE and several Republican members of Congress, particularly in the House of Representatives,[21] proposed bills suggesting changes to the Fund's spending. They wanted to accelerate the project before the DOE began to default on its contractual obligations to nuclear utilities by failing to provide an operational deep geologic repository by January 1, 1998. One of these bills was proposed in DOE's 1994 budget request and aimed to move the Fund to a "mandatory spending" format, without congressional appropriation, "to ensure that sufficient funds are made available for the disposal program" (US Government Accounting Office 1993).

In its report on the proposal, the GAO noted that the Department's proposed legislation would make more funds available to achieve the program's goals. However, the GAO's assessment presented an intriguing tension. It assessed the proposal, not only in terms of its contribution to program delivery but also in terms of the rationality with which the Fund has been imbued over time, and acknowledged that the ways in which the Fund is used "have implications for the federal deficit, congressional oversight and financial health of the program over the long-term" (US Government Accounting Office 1993). The GAO's conclusions have made the federal budget deficit one of the Fund's concerns.

All these changes led to a reconfiguration of the Fund's objectives. The political and moral reasoning behind the nuclear waste management program had to be reconciled with the political and temporal rationale of deficit reduction. At the time of its conception, the Fund had been designed as an arrangement for translating the political and moral reasoning of the nuclear waste management program into action: now, it also had to account for the moral economy introduced by Reaganomics. Measures to keep the Fund independent from and immune to political institutions' short-term political priorities were abandoned. The program was left with a Fund whose temporal, moral, and political values were constantly renegotiated. Spending on the Nuclear Waste Fund could no longer be differentiated from other annual DOE programs and entered the realm of politics.

Conclusion

In this chapter, the problem of rendez-vous to be resolved involved holding together contradicting and competing temporalities, responsibilities, and values inside a budgetary mechanism. Policymakers attempted to strive, on the one hand, to develop a stable, continuous, and sustainable budgetary mechanism that would enable the nuclear waste program to remain unaffected by short-term political changes and, on the other, to keep that very program under the control of political institutions. The reconciliation of these values was to be ensured by the adoption of a series of delicate measures, marking this initial period by the efforts to revalue, reconfigure, and reorganize one of the most traditional technologies of government: the budget.

As soon as the Fund was put in action, however, those measures were put to the test of different political actors across different budgetary authorities. This led to the instant modification of the initial script of the Fund, which transformed the nuclear waste program into an ordinary item in the federal budget, subjecting it to constant revaluation. Measures once put in place to manage and allocate the Fund became responsible for fulfilling the Department of Energy's obligations to private contractors and served the administration's short-term political priorities.

This also signaled a change in the preoccupations once expressed in resolving the problem of rendez-vous in terms of the definition given to the notion of "control." Initially, "control" was about designing a sustainable yet accountable funding mechanism. In action, it became aligned with a set of public-sector reforms launched across many OECD countries, broadly referred to as *New Public Management* (Hyndman and Connolly 2011). The objectives of these reforms were manifold: to reduce public spending and increase "the efficiency" and "effectiveness" of government actions (Barretta and Busco 2011; Pollitt and Bouckaert 2011), particularly in line with "Reaganomics." The metamorphosis of the Fund not only defined how the national nuclear waste program was budgeted but also signified a fundamental shift in the underlying long-term oriented political and moral values that once guided the nuclear waste policy, including what was to be contained—namely, an ordinary project identified in terms of its temporal, budgetary, and contractual scope, rather than the most hazardous remains of the nuclear age.

But what does all this say about institutions' capacities to design institutions to reconcile "sustainability" and "accountability"? The formulation and the career of the Nuclear Waste Fund reveal that holding together temporalities, responsibilities, and values in an institutional design is often more complex than recognized. The very categories of what should be sustained versus what should be subjected to accountability shift as definitions of values, temporal relations, and responsibilities change, often in unpredictable ways. This does not mean that efforts to reconcile these competing demands are futile. Rather, it suggests that all attempts to resolve problems of rendez-vous are temporary, partial, and experimental.

4 Death by Budget: The Abandonment of Yucca Mountain during a "Nuclear Renaissance"

> Reflecting the image of the bloodstream, public finances constitute a fluid that circulates within the state and inside the society. If the money is not often directly visible in the operations of the state, multiple practices infer and give the sign of its inevitability both for the social corps and for the political body. Briefly, the money is both widely collected and widely distributed, for numerous public policies and in the form of multiple modalities. (Bezes and Siné 2011, 21)

> For a federal agency—and for the state, local, and semiprivate organizations largely funded by the federal government—the budget is the irrigation system that provides the water without which an agency and its products would parch and wither. (Wildavsky 1988, 6)

The above quotes use biological metaphors to articulate how public finance gives life to policies and enables them, even though its role often becomes invisible. By allocating funds to certain policies and programs, a government firmly states its priorities among different public policy options. This constitutes a manifestation of its political power and the institutionalization of its political choices, equipping it with the capacity to act on what it judges to be "important," "prior," and "valuable" (Siné 2006, 319). Therefore, the constitution of a national budget does not simply involve the prioritization of "cold" budgetary figures but may also involve the lively revaluation of territories, geographies, and geologies. Budgetary action is a powerful means for concerned actors to give life to some projects while killing others, though it often operates less visibly than legislative debates or public campaigns. This chapter reconstructs the processes through which governmental and nongovernmental actors successfully brought about the death of the highly controversial Yucca Mountain project.

In chapter 3, I focused on the formulation and the early career of the Nuclear Waste Fund, a budgetary mechanism specifically designed to resolve a problem of rendez-vous that attempted to bridge the temporal gap between the budgetary times and organizational times of the nuclear waste program. Although originally designed to insulate the project from short-term politics, the way the Fund was initially legislated and the ways it transformed over time during the era of Reaganomics led to its enfoldment within political realms controlled by elected officials advancing short-term interests, including federal deficit reduction. This chapter traces the Nuclear Waste Fund's capacities to enact different political priorities from the George W. Bush administration through the Obama administration. It starts with the legislative action taken under the George W. Bush administration to revalue the Yucca Mountain project among its political priorities during its attempt to jumpstart a "nuclear renaissance." It then elucidates how, over time, budgetary platforms and procedures transformed into political sites, enabling concerned actors to both question the Yucca Mountain project's contribution to the "common good" and acquire the capacity to act upon the nuclear waste program's budgetary dimensions, which led to its "killing" through stopping the financing that had been its lifeblood.

In the 1990s, concerns surrounding climate change had been increasingly acknowledged. Carbon emissions generated by different energy sources were constantly measured and compared (Saraç-Lesavre 2015, 2021b), leading to the requalification and revaluation of nuclear energy, especially among nuclear countries with the capacity to build and export nuclear technologies as a "clean and green" energy source. In tandem with its intensive efforts to relaunch the US nuclear industry, the George W. Bush administration made the acceleration of the Yucca Mountain project a political priority.

This was greatly facilitated by the elimination of the only explicitly stated political procedure available to the State of Nevada to oppose its siting. The nuclear waste legislation provided the State of Nevada with a partial veto: a "notice of disapproval" that could be submitted upon the Department of Energy's completion of the Environmental Impact Statement, after the Secretary of Energy had recommended the site to the president for approval. Fifteen years passed between the designation of Yucca Mountain as the only candidate site and the time when the State of Nevada was finally able to use its right to exercise a partial veto. This veto was only partial since it could be overridden by a joint resolution passed by both Houses, and this is what happened.[1] In 2002, Congress passed a joint resolution overruling the State of Nevada's notice of disapproval (Stewart and Stewart 2011, 203), whereupon the president approved the Yucca Mountain project, and the Department of Energy (DOE) began preparing to submit its licence application to the Nuclear Regulatory Commission (NRC). Concerned actors, including the State of Nevada, no longer had access to political means that had previously been designed to enable them to intervene in the project's future.

In the meantime, due to a whole series of technoscientific, political, budgetary, and regulatory problems and controversies, the project was significantly delayed and behind schedule. The regulatory science produced by the scientists and engineers of the Office of Civilian Radioactive Waste Management (or the Office), as well as the regulations and standards set by the NRC, had become extremely controversial (Macfarlane 2003). Concerned scientists, ethicists, environmental groups, and anti–Yucca Mountain groups consistently pointed to various scientific uncertainties surrounding the project's ability to confine radioactivity from the biosphere for thousands of years (Shrader-Frechette 1993; Oreskes et al. 1994; Macfarlane 2003; Long and Ewing 2004; Macfarlane and Ewing 2006). They criticized the extent to which technoscientific standards, scientific methods and models, and evaluation criteria meant to assess Yucca Mountain's suitability for nuclear waste disposal were constantly being readapted in accordance with the conclusions reached by regulatory science (Macfarlane 2003). The decision as to whether or not these scientific uncertainties and controversies should lead to the abandonment of the project was left to technoscientific and regulatory procedures, seemingly independent from politics.

Debating Collective Values at the Budgetary Platforms

The absence of an operational deep geological repository was a major obstacle in the path of the George W. Bush administration's nuclear ambitions, which transformed the completion of the project into a priority for the executive branch. The second Bush administration, in particular, were the very last years when the DOE could expect the project to receive construction authorization before the presidential elections due to the NRC's lengthy licensing review process. With the DOE needing to submit its license application for a procedure that was expected to take three years, the DOE intensified its efforts to speed up the project and started requesting larger amounts in its annual budget requests. These amounts exceeded the Fund's annual revenues, and as noted in chapter 3, the Office of Management and Budget (OMB) had removed the DOE's ability to request these funds. To make the appropriation of these higher amounts possible, the DOE and the project's Republican supporters in Congress engaged in legislative efforts to liberate the Fund from both the budgetary constraints imposed by the OMB and the oversight of congressional appropriations committees.[2]

One of the proposed bills suggested changing the status of the Fund by reclassifying the fees paid into the Nuclear Waste Fund as "offsetting collections" for the fiscal years 2005–2009. If the Fund changed status and was reclassified as "offsetting collections," then spending would have become subject to the Energy and Commerce Committee authorizations and no longer to the Appropriations Committees' budget allocation for the Energy and Water bill. This was a substantial modification. Offsetting collections are not added to the receipts: they are directly credited to expenditure accounts and deducted from gross budget authority and can be spent without further congressional action. Therefore, they would not require the approval of the Appropriations Committees. By facilitating the appropriation of increased levels of funding and reducing congressional oversight (The White House n.d.), political actors supporting this bill sought to reinscribe a new relationship of control and independence in the Fund.

During the congressional hearings held on the reclassification of the Fund, one cannot help but observe the diversity of perspectives, along with the diversity of communities, geographies, and geologies that the budgetary platforms reassembled.[3] The deadline by which the federal government was to take responsibility for nuclear waste had expired (1998), and the DOE had failed to fulfill its contractual and legal engagements. Nuclear utilities had begun filing lawsuits against the DOE and receiving compensation for the extra time, during which they had to continue storing spent nuclear waste on reactor sites. Each time a court ruled against the DOE, compensation was paid from the US Treasury Judgement Fund: therefore, using taxpayer monies. The completion of the project rapidly became a priority for the administration in its effort to avoid both taxpayers and the "tens of millions of households who use nuclear-generated electricity" (and are therefore ratepayers who pay the nuclear waste fee) from being financially burdened with those overlays (Public Citizens 2002). Requesting the reclassification of the Fund, the Under-Secretary of Energy chose to speak specifically on behalf of ratepayers:

> Enactment of the Administration's legislative proposal would help rectify a long-standing inequity imposed on all of the tens of millions of households who use nuclear-generated electricity. . . . They are not being treated fairly. By law that money is being collected for one purpose and one purpose only—performance of "nuclear waste disposal activities" through the development of a repository. In actuality, much of the total that has been collected sits unused, but on a year-to-year basis, the fees not appropriated are used to offset other federal programs. . . . As far as the millions of ratepayers are concerned, the fee appears to be just one more federal tax instead of the payment to ensure the delivery of a service that it was intended to be.[4]

The Under-Secretary based his argument in defense of the liberation of the Fund from congressional oversight on the legal and contractual foundation of fee collection. To him, in a relationship of mutual legal liability,[5] ratepayers were paying a "fee" on their electricity bill, so placing the Fund under the direct authority of the DOE was to provide them with the service for which they had been and were making payments.

"Ratepayers" were materially, socioeconomically, and financially impacted by the nuclear waste program at different levels. Congressional representatives from nuclear geographies where spent nuclear fuel from decommissioned reactors was still stored, along with representatives of nuclear geographies where nuclear electricity continued to be generated and used, were particularly active in congressional debates concerning the fate of the Fund. Among those who supported the Fund's liberation from congressional oversight, the House Representative from Maine argued that the collection of the fees had lost its legitimacy, as it could not be used to fund the project effectively.[6] Maine continued to host one of the many shutdown nuclear reactors in the United States: the Maine Yankee reactor. At the reactor site, spent nuclear fuel continued to be stored, awaiting removal, an activity that not only required constant monitoring and oversight but also did not allow the land to be reused for other purposes.

Congressional representatives from Nevada problematized the issue based on neither the political and legal legitimacy of fee collection nor the federal government's contractual and legal commitments to provide services. They focused instead on the objective of the legislative proposal, which they challenged. Protesting the acceleration of a project they deemed technically and scientifically unsound and unsuitable, they argued that decreasing congressional oversight and increasing appropriation would put the "nation" at risk. Therefore, they saw the proposed changes as serving the "private interests" of the nuclear industry and burdening not only the State of Nevada, but also the "nation":

> The bill before this committee, H.R. 3981, would effectively remove the Yucca Mountain project from Congress' budgetary authority and eliminate any congressional oversight of the project for future Congresses. The passage of such legislation poses a great danger to my State, but also to the Nation. Not just the State of Nevada, but the Nation on a physical level and on a symbolic level.[7]

Through congressional debates, other environmental groups, taxpayer defense groups, anti–Yucca Mountain groups, and antinuclear groups were able to express their concerns and question both the political and technoscientific legitimacy of the project. The list of signatures attached to a common letter against the bill and the modification it would bring to the appropriations process shows the diversity of organizations and associations involved: the National Environmental Trust, the Natural Resources Defense Council, Nevada Nuclear Waste Task Force, the Nuclear Information and Resource Service, Physicians for Social Responsibility, Energy Program Public Citizen, US Public Interest Research Group, the Global Warming and Energy Program Sierra Club, and Taxpayers for Common Sense Action. Describing the project as "troubled," the letter asserted that spending more of the Nuclear Waste Fund would be "irresponsible" and remove all democratic accountability.[8]

The bill did not pass. Across their diverse stances on the Yucca Mountain project, concerned actors were able to voice their political positions on the government's current plan to store nuclear waste through the mediation of budgetary platforms. This turned the project into a more political issue, with those who were erased in the designation of the Yucca Mountain site becoming able to have their positions heard through budget appropriation platforms in Congress, which became "political sites" for debates about both the specific program and larger issues related to nuclear waste (Barry 2002).

"Cardinals" of the Senate

Up to now, I have left aside one of the most important actors in the congressional budgetary process: the Appropriations Committees, where the valuation through budget takes place annually. The failure of the legislative bills to revalue the Yucca Mountain project and to enact higher figures for the program cannot be understood in isolation from them.

Appropriations Subcommittees are very powerful congressional committees. Both the House and the Senate Appropriations Committees are composed of several subcommittees, while the complicated character of appropriations renders the subcommittees substantially independent from the parent committee (Wildavsky 1988, 18). Their chairs are called "cardinals," and being a member of an Appropriations Committee means holding an important position in Congress (Schick 2000, 200). Each "cardinal" holds considerable autonomy in running their subcommittee in a fashion that is "free from interference by the full Committee chair or party leaders. It is also the tenure on the Subcommittee, rather than on the full-committee seniority, that determines who becomes the cardinal" (Schick 2000, 200–201). Being a "cardinal" does not mean that the budgetary figure advocated by the "cardinal" would be directly adopted. The translation of budgetary authority into political power is the consequence of constant negotiations. I use the notion of "power" not as something pregiven but as an "effect" or an "outcome" (Rose and Miller 1992, 184).

With the attachment of the Nuclear Waste Fund to the DOE's annual discretionary spending, the subcommittee had gained political importance for the future of the nuclear waste program. However, its jurisdiction was not limited to the Yucca Mountain project and included, among other things, the overall DOE and the NRC budgets. During the heated debates surrounding the future of the Yucca Mountain project, Democratic Senator Harry Reid from Nevada and Republican Senator Pete Domenici from New Mexico were the "cardinals." From June 1997 until September 2006, Reid and Domenici alternatively served as the chairman of the subcommittee depending on the Senate's majority. Furthermore, from 2005 to January 2007, Reid also served as the Senate minority leader. In 2007, Reid left the subcommittee and became the Senate majority leader. In parallel, Domenici continued serving as a ranking member until he retired in 2009.[9] While the two occupied different positions politically, they engaged in various arrangements and alliances, constantly negotiating their political priorities around the budgetary figure to be enacted by the subcommittee.

Reid and Domenici each pursued their own political programs. A major proponent of nuclear power in the Senate, Domenici encouraged and sponsored legislation launching new federal nuclear programs under the George W. Bush administration. While Domenici generally supported the Yucca Mountain project, his support was far from unconditional. Back in

1992, disappointed by the Yucca Mountain project's slow progress and its increased costs, Domenici (together with Louisiana Senator J. Bennett Johnson) proposed the termination of the project to the Senate Committee on Energy and Natural Resources (Metlay 2000, 215). While this proposal failed, going forward, he would continue to operate based on concerns about the program's feasibility and operations. For his part, just before Yucca Mountain was designated as the sole site for a repository, Harry Reid had been elected as a junior senator. Opposing Yucca Mountain would become a central theme of his political career. In 1999, Reid was first appointed as the Democratic Senate minority whip (vice) leader. His political career blossomed as the George W. Bush administration accelerated the Yucca Mountain project.

Both politicians took a pragmatic approach to budgeting. For instance, during the negotiations over the DOE's 2003 budget request, Reid supported the Advanced Fuel Cycle Initiative, then newly launched by the administration, a political priority for Domenici, while Domenici and Reid cosigned the Senate bill that appropriated $155 million, less than the amount that the DOE had requested for the Yucca Mountain project (Grove 2003). In 2005, the administration passed the Energy Policy Act,[10] which promised loan guarantees to the nuclear industry for new nuclear facilities. The absence of an operating repository was considered to constitute a serious setback to a "nuclear renaissance" for the nuclear utility owners who were to invest in the construction of new nuclear reactors.

In Congress, those who opposed and supported the "nuclear revival" started to develop "third way" options. This entailed seeking alternative methods of handling nuclear waste. For instance, in 2005, Reid, together with a Republican senator from Nevada and two Republican senators from Utah, cosponsored the Spent Nuclear Fuel On-Site Storage Security Act of 2005.[11] The bill proposed that spent nuclear fuel should be transferred from spent nuclear fuel pools to dry casks to be stored in surface facilities. The Committee report asked the Department of Energy both to plan for the interim storage of spent nuclear fuel at federally operated facilities, to be located within the states where nuclear power plants operate, and to appropriate federal funding for their development.[12] Domenici supported this plan (The Associated Press 2006).

By definition, the on-site storage plan excluded Nevada, since Nevada did not have any nuclear power plants. Reid and Domenici expected entirely different finalities from the proposal. For Domenici, this proposal was a means to "clear the way for the industry to start building nuclear plants." For Reid, however, "the plan to build a string of temporary Yuccas across the country is certain to draw strong protests from the states involved, nudging them to agree with him that the waste should be kept at the plants" (*Las Vegas Sun* 2006).

The *Las Vegas Sun* newspaper described Domenici's actions during that period as those of "Dr. Jekyll and Mr. Hyde." This metaphor highlights the politically vulnerable nature of the arrangements reached between the two senators. Four months after the announcement of that plan, in September 2006, Domenici cosponsored another bill with Republican Senator Larry Craig from Idaho[13] on the Committee on Environment and Natural Resources. The bill proposed sending all spent nuclear fuel generated in the United States to Yucca Mountain through the removal of the project's capacity limit, which had been enacted in the NWPA Amendment of 1987. The same bill also proposed commencing shipments to Yucca Mountain by 2010. Politicians had an increased sense of urgency to do something, whether it was to stop or accelerate the Yucca Mountain project. One can suggest that Yucca Mountain was being reimagined as not *the* primary site of containment, but perhaps one in a more dispersed and varied nuclear waste disposal landscape, in which more political actors had stakes.

In a competing attempt, in March 2007, Reid cosponsored a bill with Republican Senator John Ensign from Nevada and Republican Senator Robert Bennett from Utah on the Committee on Environment and Public Works.[14] The bill proposed the continued storage of spent nuclear waste at reactor sites, which was to constitute an alternative option to its disposal at Yucca Mountain. The Yucca Mountain project was thus constantly revalued depending on the altering political priorities of political actors; meanwhile, the individual careers of some "cardinal" political figures became attached to and influential over nuclear waste policy.[15]

While holding the position of a "cardinal" was indispensable for undertaking political action through the budget, Reid's ability to intervene in the project was mainly the consequence of numerous negotiations, rather than a given budgetary "power." Jane, a former Yucca Mountain scientist and manager, characterizes the project's situation during that period with a metaphor that she and her colleagues liked using: "The sick cat of the neighbourhood, no-one takes enough care for it to heal, but no one let it starve to death either."[16]

The Temporality of Annual Budgeting as an Organizer of the Materiality of Work on Site

Over time, the work at the project site started to be organized following the temporality of the appropriation process. Jane characterizes the budgetary mechanism as the "Big Achilles' heel of the program" and describes with despair the difficulty they had in operating the program under these conditions:

> So you don't know how much money you will get right up until you get it. You can't move effectively. No company would do that. When you build a building, they know how much money it would cost and they have that money set aside in their budget, and the Congress doesn't work that way.[17]

Jane complained about how the timing of the enactment of the annual appropriation bill constantly generated anxiety and hesitation among program managers. The appropriations process takes the following form: Every year in February, the president submits a budget to Congress, launching the appropriations process. During February and March, each agency submits material that justifies its budget request, and the committees have preliminary meetings with the OMB. Between February and April, subcommittee hearings are held. In May and June, the subcommittees must allow floor consideration of appropriations bills, which is followed by each subcommittee chair issuing a "mark" recommending amounts for the account in the subcommittee's jurisdiction. Each subcommittee marks up its bill and drafts a report explaining its decision and providing guidance to affected agencies. Then, the full committee usually only makes a few changes in the bill that has been prepared by the subcommittee, leading to the House and Senate floor actions and their final passage in September or October (Schick 2000, 203). This is a process that is repeated every year and to which, as earlier noted, the Nuclear Waste Fund was quickly associated. As a result, Jane and the others had to wait for the result of the appropriations process to work:

> This is all going on between, like, May and September or October, when they are supposed to have their appropriations done. But it takes almost a year when that pass back comes and when we turn into work, and by that time you are getting the next pass back and you are starting the whole cycle again, and every year you go through the whole process instead of it being multiyear or off-budget.[18]

Particularly in the years when the annual budget had difficulty passing Congress and the federal government operated on continuous resolution, as was the case for the fiscal year 2007, the project had to be handled under the assumption that it would receive the same amount of funding as the preceding year. For Jane, this created a double-edged problem: if Congress finally allocated less funding than the previous fiscal year or less than what the Office of Civilian Radioactive Waste Management requested, some of the project-related activities would have had to be slowed down or stopped, in case the appropriated funding had already been spent:

> So, then it depends whether they pass the budget, or whether they will have a continuous resolution. Eventually they pass the budget, and we are watching carefully, and seeing that number between $0 and $800 million. . . . We are watching, looking, and meeting, and we say, "if it is this number we are going to do this" or "if it is this number we are going to do that," so we build all these scenarios. . . . This is how it works. . . . There is this big balancing act between July, August, and September. You get to the end of the year, hopefully with some money to carry over to the next year, but not too much, and definitely not deficiently. You can't have spent more than was appropriated. In October, the Congress comes with a continuous resolution and you are stuck again with "How much money do I really have? And how much I can spend? How am I gonna get through until I know what my real appropriation is?" It is very stressful.[19]

Thus, the work on site had to operate in coordination with the temporality of the appropriations process, which configured the scope of the scientific research, engineering work, and various investments related to the project. This constantly generated uncertainties about the execution of the project and prevented long-term planning. The nature of the work being done on site was changing in material ways due to this process; creating plans for a number of budgetary scenarios involved an enormous amount of labor on the project site, beyond the scope of the original plans. Thus, the budgetary process not only entailed the hierarchizing of political priorities through the figure enacted in the budget, but the temporality of the budgetary process also substantially reshaped and reorganized the scope, temporality, and materiality of work on site.

Presidential Campaigns and the Yucca Mountain Project

Soon, the future of the Yucca Mountain project became a subject of presidential campaigns. On June 3, 2008, just before the presidential elections, the DOE submitted its license application for the Yucca Mountain project to the US Nuclear Regulatory Commission (NRC), seeking authorization to construct the geologic repository at Yucca Mountain, despite the contestations of the State of Nevada.[20] The project had never been so close to becoming operational.

During the campaign for the presidential elections, the Republican and Democratic candidates used different repertoires to define the "national interest" regarding the Yucca Mountain project. Democrats denounced the project, arguing that it resulted in a democratic deficit, in that the designation of Yucca Mountain relied on a majority vote in Congress, despite the opposition of the State of Nevada, and held the potential to jeopardize the "national interest" by failing to contain radioactive waste. By contrast, McCain, the Republican candidate, argued that the project served the "national interest" by allowing for increased electricity generation from nuclear energy, enhancing "energy security" and confining the number of sites where spent nuclear fuel is stored into "one," thus contributing to "Homeland Security" (McCain 2008).

In fact, the Democratic candidate Kerry had already expressed his opposition to the Yucca Mountain project during his 2004 presidential campaign (VandeHei 2004). However, during the 2008 presidential election campaign, "swing states" became more important for the Democratic Party campaign, which pushed the project's fate higher on the electoral agenda. With no certainty of winning in Florida, the Democratic Party had put effort into winning the other swing states, including Nevada.[21] In the two preceding presidential elections, the majority of Nevadans had voted for the Republican Party. Thus, shifting Nevada toward the Democratic Party became an important objective.

During the campaign for the nomination of the Democratic presidential candidate, both Hillary Clinton and Barack Obama visited Nevada and officially announced their stance against the Yucca Mountain project. In a letter sent to the *Las Vegas Review-Journal*, Barack Obama again clearly stated his opposition to the Yucca Mountain project (Obama 2007). He redefined the collective values to be pursued by the US nuclear waste program as the enhancement of state authority and the adoption of a consent-based decision-making process that would overcome the federal/state authority distinction. Furthermore, he emphasized the necessity of restarting a process based on "sound science" and associated the Yucca Mountain project with the "private interest" of the nuclear industry. He disassociated himself from "the D.C. lobbyists or political action committees." The candidates' positions on the Yucca Mountain project thus became a differentiating factor.

When McCain and Obama won the candidature of their parties, Obama characterized the termination of the project as a way to "protect our families" (Wheaton 2008), while McCain characterized opposition to the Yucca Mountain project as something that was "harmful to the national interest" (Ralston 2008).[22] Barack Obama won the presidency, and the Democratic Party won Nevada, which transformed the termination of Yucca Mountain into a presidential promise.

As soon as the Obama administration took office, the new Secretary of Energy, Steven Chu, announced the termination of the project. President Obama asked Secretary Chu to appoint a Presidential Advisory Commission, called the "Blue Ribbon Commission on America's Nuclear Future."[23] The Secretary of Energy charged the Commission with assessing options for the nuclear waste program, while explicitly excluding Yucca Mountain from the Commission's agenda: the Commission was to look toward the future, not to "review the past."[24] In addition, the DOE submitted a request to withdraw its license application for Yucca Mountain from the Nuclear Regulatory Commission. However, until the NRC ruled on the request, the NRC had to pursue its licensing review process further. Therefore, before ruling on the request, the NRC could potentially provide the DOE with authorization to construct.

Reformulating nuclear waste policy while removing the Yucca Mountain project from the national legislation was not an easy task. This required both that the Democratic Party hold a majority in both Houses and that congressional members would be willing to change the legislation. The Democratic Party held a majority in both houses of Congress during the 110th and the 111th Congresses (between 2007 and 2010). In the 2010 mid-term elections, the Republican Party won the majority in the House, achieving their highest number of seats since the mid-1940s. While the Democratic Party kept its Senate majority, they lost seats.[25] Therefore, the administration did not have enough support to modify the NWPA Amendment of 1987. In addition, not knowing whether the NRC would accept the request to withdraw the license application meant risking a potential change in the next presidential elections, leading to the project receiving authorization to construct.

"Killing" the Yucca Mountain Project

In the Nuclear Waste Policy Act of 1982, the architects of the act placed funding structures for the nation's first nuclear waste management program in a section dedicated to "other provisions." Distinctly laid out technoscientific, political, and regulatory procedures were to determine the fate of the nuclear waste program, with the Fund acting as a supporting intermediary to enable its execution over a long period of time. In contrast with the rather low level of importance it was originally accorded, in late July 2009, Reid announced that he had reached an agreement with the White House on the level of budget to be appropriated to the Yucca Mountain project (Mascaro 2009). For the fiscal year 2011, the Secretary of Energy requested no funding for the Yucca Mountain project and only added funds for the closure of the Office (OCRWM) in its budget request.[26]

Aaron Wildavsky highlights how the congressional budgetary process typically relies on incrementalism rather than constant optimization; public bodies in charge of crafting budgets adjust the level of funding by basing their analysis on the preceding year to reduce the time and effort involved in its constitution (Wildavsky 1988). Drastic changes in the budget allocation manifest a change in political intentions and express a reorientation (Siné 2006, 319). Thus, the budgetary process involves a sort of double movement that relates the allocation of resources to the affirmation of values and constitutes a scene of "conflicting promises" that is a "fundamentally ideological" one (Wildavsky 1988, 320). This insight draws our attention to the extent to which the budgetary actions and inactions on funding Yucca Mountain had become the stage for significant ideological reorientations regarding nuclear waste disposal and problems of rendez-vous it posed, even as the mundane nature of budgeting may initially obscure this pattern.

The DOE required that the activities associated with the closure of the Office should be completed by the end of the fiscal year. Thus, the termination of a project that had been under development for almost thirty years and designed to contain nuclear waste for thousands of years was to last only a single "fiscal year," during which time its activities would be limited to shutting itself down. In the subsequent years, the DOE continuously requested zero funding for the program.[27] The negotiations between the Senate and the House were limited to determining whether or not stopping funding meant the termination of the project: the House Appropriations Subcommittee attempted several times to appropriate more funding to the program in its 2012 and the 2013 budget requests.[28] Even though the House succeeded in appropriating $25 million to the project in 2012 budget request, these funds were suppressed in the final Appropriation Acts.[29]

The abandonment of the project required the closure of the Office in charge of the Yucca Mountain project, which was primarily located in Las Vegas, Nevada. Since the management organization was not designated as an independent agency and was placed under the DOE, once the administration decided to terminate the project, the closure began with the Office, which had very limited time and budget. The sooner the project was stopped, the less likely that its closure could be reversed by regulatory, judicial, or legislative authorities.

The limited timing and budget determined the form of the decisions made by the Office to implement the closure of Yucca Mountain. The Office had two options (Office of the Inspector General 2010): to either follow a screening process or to declare the property abandoned. The screening process is lengthy. It requires the storage of the equipment, determining whether or not it should be sold, donated to another agency, or reused by the agency to which it belongs. In the alternative option, the federal agency declares the "excess property" abandoned. The Office opted for the second option. Equipment in the Las Vegas offices and at the Yucca Mountain site was disposed of by the agency: this included large containers and warehouses, railroad tracks, computers and office furniture, scientific and construction equipment, water monitoring equipment and tractors, and so on.

The Office's property was transferred to different federal sites: about twenty-five truckloads to the Hanford Site, the West Valley site in New York, the National Nuclear Security Administration in Las Vegas, and the Sandia National Laboratory in New Mexico, while some of the computer equipment was donated to the area's schools. Later, the railroad tracks outside the mountain were removed. Not all materials at the Yucca Mountain site were suitable for transfer under the abandonment regulation, as the fees paid by the ratepayers were used to finance the material. Therefore, the Office had to sell them at "fair market value" and return the income to the Nuclear Waste Fund (Office of the Inspector General 2010). For instance, the Yucca Mucker, the machine that dug the hole inside the mountain (see figure 4.1), and other materials were put up for sale at federal auctions in Las Vegas. One of the former Yucca Mountain scientists was amused by the sale of the "Yucca Mucker." To him, it would be so costly to transport the Yucca Mucker that no one would have wanted to buy it, as the purchaser was obliged to transport it at their own expense: once dismantled, ten large trucks were necessary to transport the machine.[30] In the spring of 2013, after the closure of the

Figure 4.1
The "Yucca Mucker," the tunnel-boring machine, Nye County, Nevada. *Source*: US Department of Energy.

Office, the DOE was planning to remove the remaining material inside the mountain.[31]

As with the material, the people working for the project were dispersed all around the United States (Yehle 2010). In the spring of 2012, during my fieldwork in Nevada, the office buildings that had hosted the Office were empty: the DOE had rented them for several years. The small restaurant in front of the office buildings had gone bankrupt. The federal employees were given a year and a half to find new jobs within other governmental agencies, while the contracts with external companies were terminated, and their employees were encouraged to seek new employment as soon as they could.[32] At the end of that period, the number of employees at the Office had been reduced to twenty. On the mountain, except for the huge hole that could not be approached, there were still a few accessible remains of the Yucca Mountain project. The duration of the termination—"a fiscal year"—had determined the materialization of termination.

Terminating the Project's Regulatory Existence through its Funding

Despite all the actions undertaken by Reid and the executive branch toward the termination of Yucca Mountain, the State of Nevada continued financing and hosting a task force against the project. An excerpt from an interview given by the head of the task force to a reporter from the *Las Vegas Sun* articulates the reason:

Journalist: Sen. Harry Reid, D-Nev., has proclaimed the DOE project is "dead" or at least bleeding to death from the lack of federal funds.

Head of the Task Force: Reid has vowed to watch it fail as he continues to use his leverage as Senate majority leader to slash its purse strings.

Journalist: So, why does Nevada need an Agency for Nuclear Projects?

Head of the Task Force: Because the beast is not dead yet. (Rogers 2009)[33]

For the spokesperson of the opposition to the Yucca Mountain project who worked with the State of Nevada, it was clear that his work remained as intensive as before. Keeping the project shut down constituted a full-time job.[34] Neither the budgetary measures taken to close the Office nor the project's material dismantling were sufficient to make the termination irreversible. The NWPA Amendment of 1987, designating the Yucca Mountain as the only candidate site, was not modified by Congress, and neither did the NRC rule on the DOE's request to withdraw its license application. In fact, the NRC continued its regulatory review. From the moment the DOE submitted its license application (in June 2008), the NRC's Atomic Safety and Licensing Board had three years to decide whether to license the project, thus placing a temporal constraint on the DOE's actions to terminate the project. In the meantime, proponents of the project found themselves in the position of advocating for its revival, even though it was legally still alive.

Former Yucca Mountain scientists and engineers expected the licensing review process to reach a favorable conclusion, giving the DOE construction authorization for the project.[35] Jane, a former Yucca Mountain scientist and manager, describes the termination of the regulatory procedure as follows:

> There was a process that was put in place. They said, "You study it," they said, "you take it to NRC, and you prove your case one day or not. You prove it can be a repository, or you fail, and it cannot." Then, you get a construction authorization or not, and then, the country could decide once you reach that moment. But you need to finish that process that says: "I did all the scientific work, I did all the engineering work, I put it in the license application." Whether the work we did

> was adequate or not depends on the construction authorization: there is a feeling of incompleteness, that somehow it just dropped to the whole floor. . . . And if you finished that process and the NRC said, "They succeeded in proving their case and here is the license application," but if you decide to go and do something else, lots of people wouldn't care about the Congress saying, "Thanks for the construction authorization but we are going to go and do something else." I mean that would be ok, because we proved our case. But it is not the case. It is not where we are: instead, we just stopped without closure. And that is where a lot of people come from. The problem is not that they won't build the repository, but they didn't get to take the test, to see if they passed, and everybody wants that. It is like taking a whole year to do a very difficult course or something, and then the professor says, "You know that final exam? Never mind, you don't pass, or you don't fail. You have to do it another time." There is no sense of finishing.[36]

In the absence of a new nuclear waste legislation adopted by Congress (or the modification of the existing one), the termination could not be limited to the physical termination of the project but also needed to include the termination of technoscientific and regulatory procedures underway, especially those taking place at the Atomic Safety and Licensing Board of the NRC. Otherwise, the termination had the potential to be merely temporary.

The Atomic Safety and Licensing Board is a separate entity in charge of the licensing review within the NRC. It is "an independent trial-level adjudicatory body" composed of a panel of three administrative judges, who are not all legal judges: some are technical judges who are often experts in the field of nuclear engineering (Congressional Research Service 2012). Judges are employees of the NRC, and their decisions are subject to Commission review. However, the Administrative Procedure Act, as well as the agency's traditional policy, "grants the judges independence from the Commission."[37] The Commission is made up of five commissioners, including the chairman, who is appointed by the president and confirmed by the Senate. The commissioners are each politically appointed for five years, and no more than three can come from the same political party.[38]

The DOE's request to withdraw its license application was submitted to the panel of administrative judges, who could accept or decline it. The request took the form of a "Motion to Withdraw" (Atomic Safety and Licensing Board, US Nuclear Regulatory Commission 2010b). The approval of the motion would have implied the removal of the DOE's license application from the licensing review procedure. The motion also included a clause to render the termination irreversible: if the license application was successfully withdrawn, the DOE could never resubmit a new license application concerning the Yucca Mountain project in the future.

For Jane,[39] the project's future needed to be determined by the technoscientific and regulatory procedures already underway, and the administration's direct intervention in these procedures reflected its refusal to follow democratic will, as enshrined in law by Congress. This was neither simply the personal consideration of an individual scientist nor limited to other former Yucca Mountain scientists and engineers. The Obama administration's intervention in the regulatory form of the project resulted in disagreements among various parties. As soon as the motion arrived in front of the licensing board, the board argued that such an action could only be justified if the DOE had technoscientific concerns. It stated, "Specifically, the NWPA does not give the Secretary the discretion to substitute his policy for the one established by Congress in the NWPA that, at this point, mandates progress toward a merits decision by the Nuclear Regulatory Commission on the construction permit" (Atomic Safety and Licensing Board, US Nuclear Regulatory Commission 2010a).

In June 2010, the board denied the motion. The following day, the Commission announced that it would review the board's decision by issuing an order and inviting participants to file briefs asking whether the Commission should review, reverse, or uphold the board's decision by proposing that the NRC commissioners should vote on it (Office of the Inspector General, US Nuclear Regulatory Commission 2011). Following that process, the Commission brought the board's denial of the motion to the vote of its commissioners. The vote was inconclusive. The status of the motion transformed into a complex legal and political issue, generating controversies among the NRC commissioners, between the NRC chairman and the commissioners, between the board and the Commission, and between congressional representatives. Federal courts were also solicited to decide on the status of the motion (Congressional Research Service 2012).

In this context, the administration's ability to withdraw the license application became a matter of legal and political debate at multiple levels of regulatory review. In the absence of a decision on the DOE's "Motion to Withdraw," it was once again the budgetary mechanism that enabled political action. Through the negotiations revolving around the NRC's budget, the future of the Yucca Mountain project transformed into a political football. The NRC budget is also subject to the Subcommittee on Energy and Water

Appropriations. For the 2011 fiscal year, the NRC only requested $10 million, which was $19 million less than the previous year's request (US Nuclear Regulatory Commission 2010). That amount was determined so that it would be sufficient for the "orderly closure" of the licensing review process at the NRC and insufficient for the NRC to continue its licensing review process.

However, that specific year, the president's budget failed to receive congressional approval, which rendered the political action undertaken through the budget particularly explicit (Congressional Research Service 2012). In common practice, when the president's budget fails to receive congressional approval, federal agencies are expected to follow the procedures set for continuing resolutions, which are joint resolutions to enable federal programs to continue without interruption. These are generally temporary measures and imply that federal agencies should operate at the same budget level as during the previous fiscal year (Schick 2000, 290).

In accordance with this common practice, for the 2011 fiscal year, the NRC was expected to operate at the level of the 2010 budget appropriations, which meant that the NRC could pursue its licensing review. However, in October 2010, the chief financial officer of the NRC released specific budget guidance for the NRC's activities related to the nuclear waste program and asked the NRC staff to continue their Yucca Mountain activities in accordance with the NRC's 2011 budget request (implying the closure activities) and not in accordance with the funding the agency had appropriated for the 2010 fiscal year (US Nuclear Regulatory Commission, Chief Financial Officer 2010).

The budget guidance generated disagreements within the NRC. Two Republican NRC commissioners formally opposed the guidance, as it overruled the board's ruling (Office of the Inspector General, US Nuclear Regulatory Commission 2011; US Government Accountability Office 2011a, 2011b, 2011c, 2012). Moreover, the neutrality of the NRC chairman was questioned. The House Republicans who gained a majority in the November 2010 midterm elections accused the NRC's chairman of being biased vis-à-vis Yucca Mountain–related actions,[40] particularly as he had previously served as the scientific advisor in the office of Senator Harry Reid.[41]

Following several allegations and complaints, the NRC's Office of the Inspector General (OIG) initiated an inspection of the NRC's closure activities concerning the Yucca Mountain project. This inspection particularly focused on the chairman, who was alleged to have intentionally prevented the

Commission from completing the ruling on the board's decision to deny the motion (Office of the Inspector General, US Nuclear Regulatory Commission 2011). The OIG's inspection concluded that the Office of Management and Budget, therefore, the executive branch of the government, had intervened in the reduction of the amounts that the NRC requested from Congress both for the 2010 and the 2011 fiscal years. It also noted that the chairman did not circulate information equally among the commissioners, particularly in relation to the budgetary guidance, but did not describe the chairman's actions as illegal.

In September 2011, the NRC commissioners sustained the board's decision to deny the DOE's license withdrawal motion. However, the NRC halted further consideration of the license application because of "budgetary limitations." Once more, the budget mediated the revaluation of the US nuclear waste program. This time, the budgetary action produced a displacement in a procedure that was explicitly designed to operate "independently" of political institutions.

Maintaining the Project Terminated through the Budget

The figure enacted in the budget acted as a means to sustain and justify the NRC's inaction in front of judicial bodies. Both private nuclear utilities whose spent nuclear fuel storage systems (mainly spent nuclear fuel pools) were reaching the limits of their capacity and states that demanded the expedited cleaning up of their territories from nuclear waste began submitting a myriad of lawsuits to the federal courts against the Motion to Withdraw, first against the DOE and later against the NRC.[42] Through these lawsuits, plaintiffs asked federal courts to order the NRC to comply with the law and to continue processing the Department of Energy's license application for the Yucca Mountain project.[43] The plaintiffs included the states of Washington and South Carolina, in which two of the three major federal facilities storing high-level waste produced for military purposes are located, as well as Aiken County in South Carolina,[44] the Prairie Island Indian community in Minnesota,[45] and the National Association of Regulatory and Utility Commissioners (NARUC),[46] some of whom were also among the actors who had advocated for the liberation of the Fund from congressional oversight.

By 2012, following the recommendations of the Blue Ribbon Commission, the DOE started developing a new nuclear waste program, and congressional members were working on legislative proposals to reformulate the nuclear waste policy. The Yucca Mountain project was materially terminated. Meanwhile, the US Court of Appeals ruled on the lawsuits filed against the NRC and the DOE concerning the procedures they followed for the termination of the Yucca Mountain project.

In May 2012, the US Court of Appeals ruled against the Motion to Withdraw and asked the NRC to pursue the licensing review of the Yucca Mountain project, concluding that the NRC had violated federal law by declining to process the license application. This was expected to allow time for Congress to clarify the situation by modifying the legislation defining its nuclear waste policy.[47] For the court, this was the only means through which the current status of the Yucca Mountain project could become legally justifiable. In response to the successive rulings of the court, the NRC defended its inaction by citing its lack of funding and suggested that its inaction was a service to taxpayers—a contribution to the common good—and that the lack of funding reflected Congress's unwillingness to pursue the Yucca Mountain project. In response to this justification, in August 2013, the court took a formal position on the NRC's inaction:

> Congress speaks through the laws it enacts. No law states that the Commission should decline to spend previously appropriated funds on the licensing process. No law states that the Commission should shut down the licensing process. And the fact that Congress hasn't yet made additional appropriations over the existing $11.1 million available to the Commission to continue the licensing process tells us nothing definitive about what a future Congress may do.[48]

The Court of Appeals ruled that by looking at the level of funding it expected from Congress before undertaking action, the NRC has been reading the political intentions of Congress, and by doing so, it has been replacing the role of the legislative branch of the government. Therefore, it has been in breach of the principle of the separation of powers. In its final ruling, in August 2013, the court stated,

> This case has serious implications for our constitutional structure. It is no over statement to say that our constitutional system of separation of powers would be significantly altered if we were to allow executive and independent agencies to disregard federal law in the manner asserted in this case by the Nuclear Regulatory Commission. . . . But unless and until Congress authoritatively says otherwise or there are no appropriated funds remaining, the Nuclear Regulatory Commission must promptly continue with the legally mandated licensing process.[49]

Through figures enacted in the budget provoking the NRC's actions and inactions, the Yucca Mountain project was revalued. Those figures not only enabled the constant re-hierarchization of the Yucca Mountain project among other federal programs but also enabled political intervention on procedures that were delegated to different sets of knowledges, expertise, and institutions. Although action through the budget did not permanently replace or substitute legislative, regulatory, or technoscientific authorities, it did so "temporarily"—which was long enough to produce the possibility of stopping this project that had been tightly inscribed in the state apparatus.

The court's ruling ordering the NRC to continue its licensing review was particularly interesting for what it says about this problem of rendez-vous in two ways. First, it explicitly points out how, once all the measures originally designed to render the Nuclear Waste Fund immune to short-term politics were abandoned, the budget became a political means for undertaking political action: intervening in, and even substituting for, the various branches of government. For a project expected to be completed over a hundred years and last a million years, this was an unexpected outcome. Second, this failure to keep the nuclear waste program insulated from political alterations gave the executive branch the ability to determine the fate of a project it did not want to pursue while giving concerned actors the ability to voice their opposition and act upon the project through congressional platforms.

Conclusion

As soon as the Fund was put into action, it began to leak. Since its inception, for the State of Nevada and the communities that were to host radioactive matter, a nuclear waste program that was au rendez-vous of their needs and demands meant the abandonment of the Yucca Mountain project so that they were freed from the burden of having to stay with radioactive matter deep in their underground for almost eternity. Over time, the Yucca Mountain project was also hardly considered au rendez-vous by its initial supporters since it was unable to meet their needs and demands, whether those were private nuclear utilities, communities that continued to live with nuclear remains, taxpayers, or the administration that recently took office.

The incapacity of the Fund to remain impermeable to short-term politics opened it to political debate and transformed budgetary platforms into alternative sites of politics. The Fund, once specifically designed to confine, immunize, and close the program to short-term politics, vetted a political geography that initially lacked means to oppose the siting of a project within its frontiers with the capacity to have a say in determining its future and transformed its political leaders into ones that counted in the policy debate.

Around the figures enacted in the budget, different branches of the government asserted their priorities. For the judiciary, this meant assessing whether or not procedures set by the legislative branch of government were followed. For the legislative branch, it involved constantly revaluing the nuclear waste program among a wide range of competing federal projects and programs. For the executive branch, this meant determining whether the project remained "good" and "desirable," as well as an opportunity to gain "control" over the aspects of the nuclear waste program that initially fell under the authority of other branches of the government. Finally, throughout the Fund's career, it was not only the Yucca Mountain project but also the federal government and its political will to take long-term policy action that underwent revaluation, along with the definition given to what constitutes "good" and "desirable" relations across time. If the Nuclear Waste Fund had remained loyal to its initial script, the Yucca Mountain's capacity to be au rendez-vous of the contemporary US society might still be debated at budgetary platforms, but political actors implicated in those platforms might have lacked the capacity to produce the political power that enabled its termination before the project ever became operational. In a certain way, the incapacity of the Nuclear Waste Fund, as a resolution of the problem of rendez-vous, to hold conflicting and competing values together enabled a whole set of communities, geographies, and geologies to be au rendez-vous of their ever-evolving demands and needs that had long been discounted.

III Undertaking the Nation's Nuclear Waste Futures in Enterprise Form

5 Valuing Nuclear Waste Futures: The Making of America's Nuclear Corridor

"The Town that Wants America's Worst Atomic Waste" is how *Forbes* magazine introduced the city of Carlsbad, New Mexico, in a feature article in January 2012 (Helman 2012). The title highlighted the unexpectedness of this "desire" by linking the verb "want" with the adjective "worst." As soon as the Yucca Mountain project, the only geological repository project under regulatory review for high-level waste disposal in the United States, was abandoned, a group of local actors in Carlsbad publicly volunteered their city to replace the project. The city of Carlsbad already hosts the Waste Isolation Pilot Plant (WIPP), a geological repository and a specific category of military nuclear waste[1] designated for a 10,000-year containment.

In this chapter and the next, I examine the entrepreneurial efforts of these local actors to keep their community, geology, and geography attached to nuclear futures. I trace how they attempted to align the resolution of their problem of rendez-vous, that of remaining valuable and desirable on the Earth's surface, with the resolutions of the problem of rendez-vous guided by the treatment of nuclear waste, on the one hand, as a collective concern that requires protecting the present and long-term futures from its environmental and health effects and, on the other, as a market concern whose consequences must be delineated and dealt with through the mechanisms of the US neoliberal economy.

A growing body of scholarship addresses how communities and geographies have been dealing with the remains of the nuclear age. Nuclear landscapes, it is often emphasized, entail sacrifice and dispossession—particularly of Indigenous lands, rural communities, and communities of color who disproportionately bear the health and environmental costs; radioactivity is overwhelmingly present in individuals' daily lives while remaining highly invisible to scrutiny (Davies and Polese 2015; Voyles 2015; Stawkowski 2016; Pitkanen and Farish 2017; Barad 2019; Brown 2019).

Many studies have highlighted how nuclear landscapes build on and generate environmental, social, economic, and political injustices—from elevated cancer rates and contaminated water supplies to forced relocations and the appropriation of ancestral territories (Ishiyama 2003; Endres 2009; Kuchinskaya 2014; Ishiyama and Tallbear 2022); impact bodies through radiological exposure and intergenerational health effects (Petryna 2013; Cram 2015; Brown 2016); and produce zones that exclude various forms of life through contamination, safety and security restrictions, and ecological damage (Galison and Moss 2015; Cram 2016; Alexis-Martin and Davies 2017) and result in new experiences of living (Houdart 2017).

Some of this scholarship has shed light on the efforts of local actors and communities to avoid having waste buried in their midst, efforts that sometimes intersect with broader movements that have opposed nuclear production and its colonial legacies (Kuletz 1998; Endres 2009, 2012; Houston 2013; Genay 2019; Gomez 2022). Others have detailed the coproduction of technoscientific and political orders (Gusterson 1998; Masco 2006; Hecht 2011, 2012). Yet not much attention has been paid to exploring how and why some communities and their members actively lobby to host hazardous materials despite their danger. In other words, how and why do communities come to construe remaining attached to nuclear waste futures as "valuable" and "desirable"?

To understand the processes through which some individuals and communities come to desire nuclear waste, it is useful first to foreground "being nuclear" as a relationship of and route to power and control. I take hold of the concept of nuclearity from Gabrielle Hecht, who uses it to denote how "being nuclear" is a historically and geographically contingent phenomenon that shapes and is shaped by a technopolitical order and is thus much more than simply a measurement of radioactivity on a Geiger counter (Hecht 2006). Hecht argues, and I agree, that "nuclearity signifies power"; "its absence signifies subjugation" (Hecht 2012, 22). However, while nuclearity produces an almost eternal legacy, the power associated with it does not last unless it is actively maintained. In the case of Carlsbad, a group of local actors attempted to maintain their city's "nuclear" status by inviting more nuclear waste within its boundaries, engaging in a series of valuation efforts. Through this process, they revisit previous formulations of and resolutions to the problem of rendez-vous posed by the terms under which nuclear waste came to be physically confined in their deep underground and propose alternative ones.

I take the actions undertaken by a set of nuclear advocates in Carlsbad as an opportunity to explore political and moral values around which nuclear futures are undertaken at specific sites. Continuing to construe "values" as resultant of valuation processes, I draw on John Dewey, who pointed at the peculiar relationship between valuation and desire. In his *Theory of Valuation*, Dewey states that "valuation involves desiring." Specifically, "valuations in the sense of prizing and caring for only occur when it is necessary to bring something into existence, which is lacking or to conserve in existence something, which is menaced by outside conditions" (Dewey 1939, 16).

Here, I draw on the work of anthropologists who suggest that local ethnographic accounts about the subjective transformations of individuals can enrich our understanding of how specific political and moral rationalities are enacted in practice and circulated (Binkley 2009; Fridman 2010, 2014; Brady 2014; Ballestero et al. 2023). As Daniel Fridman explains, "Ethnography offers the advantage of getting closer to subjects and collective practices, enabling [us] to understand the rationale of the enterprise in which they have embarked on their own terms" (Fridman 2014, 96). More broadly, my telling of this story is a plea for scholars who write about those who contest the siting of infrastructure projects to pay symmetrical attention to the perspectives of those who actively desire their presence. Their voices should not displace accounts of those oppositional voices whose dissent and discomfort have been rendered invisible and systematically silenced (Kuletz 1998; Zonabend 2007; Voyles 2015; Genay 2019; Gomez 2022), but their perspectives are necessary to understand how political and moral values are constituted at specific sites, whether in the context of nuclear geographies or other geographies of risk.

WIPP as Nuclear Status Symbol

The WIPP facility is only one of several US weapons complex facilities in New Mexico, which created new industries in "mostly rural, nonindustrial locations" across the desert in the Southwest (Masco 2006, 311). New Mexico's role in this economy, however, is unique. Among several other scholars,[2] anthropologist Joseph Masco describes it as the only state supporting the US nuclear economy from "cradle to grave," thereby intertwining local ecologies and communities with the US nuclear project (Masco 2006, 311).

The state's nuclear landscape began with the Manhattan Project's establishment of Los Alamos Laboratory in 1943, followed by the Trinity test in 1945—the world's first atomic detonation, which released radioactive fallout throughout the region without warning nearby residents (Masco 2006; Voyles 2015). The nuclear complex expanded further with the establishment of Sandia Laboratory in Albuquerque in 1948, which became responsible for nuclear weapons engineering and testing, cementing New Mexico's central role in the nation's nuclear weapons complex. This growing nuclear establishment was amplified with the beginning of intensive uranium extraction across the state. Subsequent decades brought uranium mining that produced millions of tons of ore, leaving hundreds of abandoned mine sites predominantly on Navajo lands, while the 1979 Church Rock spill released over ninety million gallons of radioactive waste into waterways flowing through Indigenous communities (Voyles 2015; Genay 2019).

This layered history of nuclear extraction, production, testing, and disposal creates the context within which Carlsbad's nuclear advocates operated—not as an isolated case, but as part of a broader picture where nuclear activities have created complex relationships of local livelihoods, health concerns, political identities, and contestation. From the perspective of Carlsbad's local actors, however, this nuclear landscape represented not burden or sacrifice, but opportunity and prosperity from which they felt excluded. They often narrated how northern New Mexico's nuclear facilities brought prestige, high-paying jobs, and federal investment to communities like Los Alamos, while southeastern New Mexico remained peripheral to those developments. To them, nuclear activities represented the path to economic stability and national significance that their community and region were missing. This sense of being left out of New Mexico's nuclear prosperity shaped how they approached WIPP—not as a hazardous imposition, but as a long-awaited opportunity to join the state's nuclear economy and identity—and how they constructed their identities as legitimate nuclear advocates. Their sense of exclusion from nuclear prosperity became central to how they formulated their own problem of rendez-vous—that of becoming and remaining politically, morally, and economically "valuable" in a region defined by its relationship to the US nuclear complex.

WIPP's emergence as part of this nuclear landscape occurred during a crucial transition. In the post–Cold War period, as Masco's work has emphasized,

Figure 5.1
Limited-edition silver coin commemorating the ten thousandth waste shipment to WIPP in 2011, photographed by the author.

the Department of Energy's (DOE's) major mission has become to clean up territories, in effect transitioning them from the "sovereignty of nuclear waste" to the "national sovereignty" of the federal government—a project that requires colossal amounts of federal funding (Masco 2006, 144). WIPP has been an integral part of that mission and, in a certain way, retroactively became a facility of Cold War history. This role has been the subject of formal recognition: as, for instance, the day that the DOE celebrated its ten thousandth waste shipment to the facility. On October 24, 2011, the facility's primary private contractor issued a limited-edition silver coin, highlighting WIPP's contribution to that mission (figure 5.1).

Since 1999, WIPP has been in operation, receiving regular shipments of military transuranic waste: a by-product of weapons production and nuclear research, composed of protective gear, tools, residue, debris, and other items contaminated with radioactive elements (mainly plutonium). These human-made elements have atomic numbers higher than uranium. At the time of my fieldwork, two categories of transuranic waste were disposed of at WIPP following procedures that had been established since operations began. Contact-handled (CH) transuranic waste can be handled by workers under controlled conditions without any shielding other than the containers themselves. CH transuranic waste makes up approximately 96 percent of the total volume of waste disposed of at WIPP. The remaining 4 percent is remote-handled (RH) transuranic waste, which emits more penetrating radiation than CH transuranic waste and must be handled and transported in lead-shielded casks. RH waste is disposed of in horizontal bore holes drilled into the walls of underground disposal rooms. Canisters containing CH waste are stacked in columns on the disposal room floor.

Through the mediation of WIPP, nuclear waste is disposed of inside the Salado Formation of the Permian Basin, a 250-million-year-old geological salt formation. Salt formations at WIPP were deposited in thick beds during the evaporation of an ancient ocean: the Permian Sea. These geologic formations consist mainly of sodium chloride rock. The primary salt formation containing the WIPP repository is about 2,000 feet thick, beginning 850 feet below the surface. Over time, high pressure on the solid formation would cause the salt to "creep," filling in the voids of the disposal rooms and sealing waste from the environment. While WIPP's designers expected the Salado Formation to contain radioactivity for ten thousand years, they assigned the city of Carlsbad a more temporary role in achieving that objective. Both the humans and the human-engineered infrastructure on the surface were construed as temporary interfaces to mediate the relations between human time and space, as well as geological time and space.

This temporal contradiction became particularly acute by the early twenty-first century, when Carlsbad depended on a nuclear economy, and local actors were not ready to give that up.[3] The temporally fragile state of their nuclear livelihood, in striking contrast with the toxic temporality of waste, shaped the intensity of the undertakings that followed. During my fieldwork in Carlsbad, the rate of shipments of transuranic waste to WIPP had begun to decrease. If this decline in the rate of shipment was sustained, then shipments were expected to cease completely in the 2030s.[4] In 2011, the first layoffs at WIPP took place. The number of employees decreased from the previous year's peak of 740 to 664, leaving locals anxious about their capacity to keep Carlsbad "nuclear," which they saw as threatening their livelihood. By 2012, as WIPP began creeping toward the end of its mission, local actors sought to avoid being abandoned and left behind, and they wanted to act while they could still capitalize[5] on their existing economy, geography, and infrastructure.

To understand their response, it is crucial to recognize how WIPP had transformed Carlsbad's relationship to space and power. WIPP, designed to contain nuclear waste deep inside the Salado Formation through its withdrawal from the present and future use of human civilization for the next ten thousand years, has created a "nuclear corridor," a term that local actors used to characterize their geography. To dispose of nuclear waste in the depths of the Salado Formation, governmental institutions have connected this place to the "nation" and, more specifically, to the Cold War nuclear complex and its legacy: linking Carlsbad to many DOE sites across the United States that have long been waiting for military waste to be removed from their territories. Transportation routes have been reinforced; a bypass route was created in Santa Fe; new canisters, containers, and trucks were designed; a controlled shipment system was put in place; and 25,000 people were trained for emergency planning and preparedness along the transportation routes across twenty-two US states. The Department of Energy and National Laboratories (Sandia and Los Alamos) have opened field offices in Carlsbad. The Carlsbad field office of the Los Alamos National Laboratory developed technical and scientific systems to produce knowledge about nuclear waste and the role that Carlsbad's geology might play in its containment.

This nuclear infrastructure, knowledge, workforce, and economy have also attracted private nuclear investment to southeastern New Mexico. Several private nuclear facilities have opened, including an uranium enrichment plant and an isotopes plant, which became operational in neighboring Lea County, while a low-level waste repository became operational in the adjacent state of Texas. The future of the local nuclear livelihood also relied on several political elements: a state governor who did not oppose their nuclear endeavors and a federal government in need of a "success story" following the abandonment of its only geological repository project for high-level waste.[6]

For the group of local actors who are the object of this analysis, therefore, the problem of rendez-vous to be resolved involved avoiding the transformation of their community, geology, and geography into postindustrial ruins.[7] Their intense undertakings involved attempts to sustain and extend the isolation of radioactive matter from humans and the environment around WIPP. This relied on their own capacity to translate their geology, geography, and community into a "solution" to the nation's highly distressed nuclear waste policy. They put this translation into action through *proactive valuation* efforts: a group of actors' attempts to generate new nuclear attachments and to prolong existing ones through the succession of a series of mediated and equipped valuation undertakings. In the case of Carlsbad, it was a *Nuclear Task Force* that engaged in proactive valuation.

The Mayor's Nuclear Task Force

The city of Carlsbad, the seat of Eddy County and its largest community, is located in the Pecos River Valley in southeastern New Mexico, about 25 miles north of the Texas border. The community is east of the foothills of the Guadalupe Mountains and at the northern and eastern edge of the Chihuahua Desert. Driving to Carlsbad from El Paso, one comes across several small oil drills. In the 2010s, locals regarded the oil and gas industries as unstable because the quantity of drilling relies on the fluctuating market price of oil and gas. Potash mining also constituted an important part of Carlsbad's economy. Aside from the moment when the local population decreased following the decline of potash mining activity in the late 1960s, culminating in the closure of US Borax's Carlsbad potash mine in 1967, the city's population has mostly been stable for decades. The community's relation with the US nuclear history was not limited to the WIPP. In 1961, Carlsbad was host to the GNOME test, the first nuclear test of Project Plowshare designed to explore the use of underground nuclear detonations for geographical engineering purposes (Szasz 1998; Richter 2017). The test failed to demonstrate that such an explosion could be fully contained by the geological formation but local actors proudly referred to it, since it situated Carlsbad on the nuclear map for the first time and demonstrated its potential to become a nuclear community.

During my ethnographic fieldwork in 2012, WIPP's operator, the URS (United Research Services) Corporation, was the town's major employer (City of Carlsbad 2012), with several hundred workers. URS also hired dozens of subcontractors for specific tasks. City officials proudly asserted that the unemployment rate was the lowest in the state of New Mexico (hovering around 4 percent) and that the city had the third-highest gross domestic product in the state.

I arrived in Carlsbad in the spring of 2012, specifically to observe the Nuclear Task Force's attempts to recalibrate the city's relationship to the changing political economy of nuclear waste. All local actors knew a "social scientist" was arriving in town. They tried to make sense of my presence throughout my visit. My interview requests led them to reflect on whether they should talk to me. I was not American. But more importantly, I was a woman affiliated with both a French and an American university. This led to confusion and some suspicion. For instance, before agreeing to a meeting, one of the Task Force members conducted "background research" on me, which included him attempting to identify my father on Google, whom he deduced to be a geology professor in my hometown. I often needed to negotiate my identity as a social scientist, a category that locals thought might also encompass an identity as an antinuclear activist, a French spy, or, perhaps worst of all, a social scientist who would measure their level of "acceptance" of nuclear waste. Ultimately, I negotiated my identity as an actor to whom they wanted to demonstrate the legitimacy of their nuclear endeavors and followed local activities as closely as I could.[8]

Before arriving in Carlsbad, I needed to obtain security clearance both to visit WIPP's underground areas and to enter the DOE's Carlsbad field office. I was offered, and accepted, an opportunity to stay at the house of a senior DOE Carlsbad field office manager and his family. It was my key interlocutor, Abe,[9] whom I met during an Organisation for Economic Cooperation and Development (OECD) working group meeting, who introduced me to the family. Abe was a well-respected senior manager at the DOE Carlsbad field office, who had worked for twenty-five years as a senior scientist on the Yucca Mountain project. When that project terminated, he—like so many others—had to find a new job elsewhere. His family remained in Nevada, and so he commuted to and from there frequently.

One of the most important moments of my fieldwork was when I first met Tom, a senior scientist working at one of New Mexico's two National Laboratories' field offices. Tom was eager to set the scene for me and underline his hybrid identity by introducing himself as both a "Carlsbad native" and a "scientist."[10] His claim of hybridity translated into his manner and style. He wore cowboy boots and chewed tobacco. He combined his local look with a technical vocabulary. He narrated his life trajectory around potash mining and waste disposal: "I was a miner, I knew how to dynamite!" He elaborated, "Potash mining is a family business, my father, grandparents were in that business." When he came back from university for summer holidays, he was asked to do "precise dynamiting work." He was recruited by a laboratory director who was looking for someone "who knew both how to mine and to do the science." By employing examples of specific technologies developed by the laboratory he worked for, he rendered his personal and institutional contribution to the accelerated cleanup of nuclear waste incontestable. Indeed, he considered it a moral obligation.

WIPP's decades-long presence in Carlsbad had produced many hybridized individuals like Tom, who was an important member of the mayor's Nuclear Task Force. The Task Force comprised a group of residents who were engaged in efforts to bring more nuclear facilities to Carlsbad. The residents were quite homogeneous in terms of their socioeconomic status: most were either local business owners or managers and technoscientists working at the local field offices of either Los Alamos or Sandia National Laboratories. The membership of the Task Force also included a few scientists and managers working at the DOE's Carlsbad field office, former and current city- and state-level politicians, and a senior manager at the DOE's principal contracting company for WIPP. Several city employees, a priest from a local church, and the mayor's communications consultant rounded out the membership. The Task Force was led by John Heaton, a former New Mexico state representative, who had been involved with WIPP since the 1970s.

The Task Force met every Tuesday at 07:00 A.M., before the workday started. To keep both Task Force members and the public informed of its operations, the mayor's communications consultant produced a monthly newsletter called the *Nuclear Nexus*, which presented the evolution of local nuclear endeavours with a positive and enthusiastic tone. My presence in town made it into an article titled "Fulbright Scholar Wraps Up Carlsbad Visit" (The Carlsbad Mayor's Nuclear Task Force 2012). Members participated on a voluntary basis, in part to circumvent a restriction on federal employees lobbying the federal government. When federal employees participated in Task Force meetings or events, they did so as "local citizens" rather than as representatives of their employer.[11] These restrictions also applied to the contractors. The sleight of hand of breakfast meetings therefore allowed everyone to participate in the Task Force's work.

These hybridized individuals on the Task Force constantly negotiated their identity as technoscientists, federal officials, and "Carlsbad natives" or local citizens, a phenomenon that transformed the Task Force into an ambivalent assemblage, a porous entity. The then-current mayor, for instance, who also worked for a potash mining company, often emphasized that he was a "Carlsbad native."[12] The Task Force's membership also included the former mayor, who had worked to bring WIPP to Carlsbad and was a major investor in a number of important amenities in the city, including several local restaurants and hotels.[13] This pattern of claiming local identity was widespread among Task Force members who worked at technoscientific institutions related to WIPP. Of course, their definition of "nativeness" excluded the Indigenous peoples who have been living on those lands for thousands of years and associated it with the arrival of settlers. This settler colonial construction of "nativeness"—which erases thousands of years of Indigenous presence—became central to how Task Force members legitimized their claims to the land and its nuclear futures.

Given that the presence of WIPP had fashioned so many hybrid identities, I found it difficult to reduce Task Force members to their institutional or professional affiliations. From the mayor's perspective, the Task Force would need to remain aligned around common objectives to be successful. This was not as easy as it first appeared to me as an outsider. Each potential project brought with it major potential ramifications for each Task Force member's personal and professional lives. Any given project would require different levels or types of scientific and/or engineering expertise, might result in more contracts for one laboratory than the other, and might embed Carlsbad with one nuclear waste network to the exclusion of another one. During meetings, actors disputed and realigned their priorities so that the Task Force remained the unique "spokesperson" (Callon 1986). Each member acted as a cross-spokesperson, allowing them to shift between their identities easily.

Situating Carlsbad at the Heart of the Atom

In 2012, Carlsbad's Nuclear Task Force sent a strong delegation to the International Waste Management Conference (WM) in Phoenix, Arizona. This annual international nuclear waste conference is one of the largest of its kind in the world, typically bringing together thousands of participants over five days to discuss all things nuclear. The conference's participants promote their "products and services" inside a huge exhibition hall, while experts and policymakers slip in and out of topical sessions in parallel time slots.

That year, the conference took place just after the United States' Blue Ribbon Commission on America's Nuclear Future had released its long-awaited final report on the future of US nuclear waste policy. The report concluded that "consent-based decision-making" is a sine qua non for the siting of future repositories and mobilized WIPP as an example to illustrate how such a process had been successfully accomplished in the United States. The

Commission's depiction of WIPP transformed Carlsbad from an invisible, remote, politically and legally "abnormal" facility of the US nuclear complex (Downey 1985) into a "success story" and an "example for the nation"[14] intimately related to, and inspired by, the "participatory turn"[15] in repository siting (Simmons and Bickerstaff 2006; Blowers and Sundqvist 2010; Lehtonen 2010; Sundqvist and Elam 2010; Bergmans et al. 2015). In the words of the Commission, consent-based decision-making required a potential host jurisdiction not only to show its willingness to host a repository but also to actively engage in the long and laborious process of negotiation[16] until an agreement is reached among a whole network of actors involved in the process (Blue Ribbon Commission on America's Nuclear Future 2012).

The other attendees and I witnessed the immediate consequences of Carlsbad's transformation into a "success story." The Task Force had hired a booth in the conference's exhibition hall; it had also secured a presence on the conference program with a listing just before the Cartwright Engineers, a "bronze sponsor," and just after Cabrera Services, a company providing radiological and environmental expertise. The conference's final and the best-attended panel session, moreover, highlighted the Commission's findings and, therefore, Carlsbad's success. This panel featured high-profile figures, including the Assistant Secretary of Energy, an NRC commissioner, a former Commission member, and an executive from a major nuclear utility. Carlsbad's former mayor was also on stage; his presence offered living proof that the United States maintained an operating deep geological repository.

Encouraged by the Blue Ribbon Commission's conclusions, "responsible" individuals and communities (Rose 1996; Trundle 2011) demonstrated their interest in playing key roles in the nation's nuclear waste futures before a new national framework of laws and rights was put in place. Following the conference, several local communities sent confidential letters to the DOE asking the agency to investigate the potential of their local geologies for nuclear waste storage.[17] While the Task Force members were grateful for the praise, WIPP's new visibility also made them nervous. They did not want Carlsbad to be transformed into a replicable example whose "value" could be "reverse engineered" (Paxson 2010). Rather, they wanted their site to be recognized for its "unique value for the Nation."[18] The Task Force developed a stylized map that identified Carlsbad as *America's Nuclear Corridor*, expressing the city's desire to extend nuclear attachments (see figure 5.2).

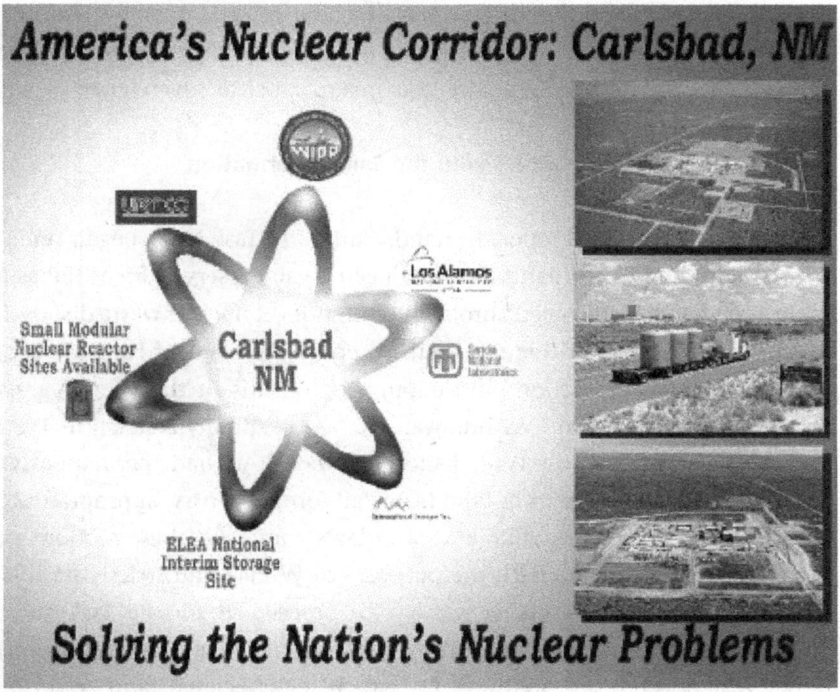

Figure 5.2
Image used as wallpaper at the Exhibition Hall, WM 2012.

In Phoenix, proactive valuation of Carlsbad's geography took a cartographical form. Cartography offers a major means for reorganizing sovereignty over a landscape; as the geographer Gary Fields writes, maps can be used "for projecting differently configured boundary lines and demarcating new areas of sovereignty and control on the landscape" (Fields 2011, 189). The Carlsbad Task Force deployed its stylized map in multiple forms, sometimes as a wallpaper at conference booths, other times as a PowerPoint slide (figure 5.2). Rather than placing Carlsbad and its region at the periphery of nuclear power, as a zone of sacrifice handling nuclear power's hazardous externalities, the map located Carlsbad at the heart of the atom, an atom that was evacuated from radioactivity, hazards, and other contestations. The map's left side displayed the already existing and potential future private and federal nuclear facilities located in and around Carlsbad. The right side showed transportation routes linking WIPP to various DOE facilities in need of being cleaned up. The map was designed as a text open to the interpretation of the audiences it was presented to, enabling them to imagine *America's Nuclear Corridor* as a potential area for nuclear sovereignty.

Aligning Nuclear Geographies with the Salado Formation

With the Commission's report in hand, Carlsbad's Task Force began pitching the specific piece of land that had been legally reserved for WIPP as a "solution" for the "unsolved" problem of high-level nuclear waste disposal. Making this case would require them to enlist the State of New Mexico, its congressional delegation, and enough politicians in the US Congress to modify the WIPP Land Withdrawal Act.[19] As originally enacted in 1992 and amended in 1996, the WIPP Land Withdrawal Act had "permanently" withdrawn sixteen sections of land from "all forms of entry, appropriation, and disposal under the public and land laws"; reserved these sections of land "for uses associated with the purposes of WIPP"; and strictly prohibited the disposal of high-level waste.[20] To proceed, then, the Task Force would have to transform the existing political and regulatory conditions under which WIPP operated and for which its associated land could be used. Only once all these transformations were completed could Carlsbad's resolution of its own problem of rendez-vous, of remaining "nuclear" and therefore "valuable" on Earth's surface, be aligned with the federal government's resolution of the problem of rendez-vous, of containing high-level waste deep underground.

The Task Force had a long list of actors whose support it needed for its project of modifying the act. To that end, its members sought influential posts within political organizations associated with nuclear geographies of high-level defense waste. One such organization was the Energy Communities Alliance (ECA), which represents "local governments that are adjacent to or impacted by the DOE activities."[21] Its Board is composed of elected officials and administrators from local governments impacted by the DOE or former DOE activities and who identify their mission as advocating "community interests." The ECA typically weighs in on nuclear waste policy by submitting letters to the Secretary of Energy and providing written and oral testimonies at US Congress concerning legislative proposals. In 2011, Roxanne Lara, who was an Eddy County commissioner, a lawyer, and a Carlsbad resident, became part of the ECA's Board through her election as secretary.[22] Members of the Task Force saw her election as a major milestone in advancing their efforts to position Carlsbad as a "solution" to the problem of nuclear waste disposal.

The ECA was only one of a series of organizations that the members of the Task Force joined. For instance, several elected officials gained entry to the National Conference of State Legislatures (NCSL), a nonpartisan organization that brings together state legislators "to support, defend, and strengthen" states' interests with regard to actions that concern them at the US Congress.[23] In 2010, John Heaton, then a New Mexico state representative and head of the Task Force, was elected chairman of the Conference's High-Level Waste Working Group. The Working Group was established following the formation of the Blue Ribbon Commission in January 2010.[24] In July 2010, he testified before the Commission on behalf of the NCSL, where he praised both the success of WIPP and "the usefulness of salt deposits" for geological disposal.[25]

While these organizations were important for disseminating the Task Force's "solution," the group also wanted more direct access to Capitol Hill. The City of Carlsbad and Eddy County therefore launched a bid for a lobbying firm to represent their interests in Washington, DC. The idea was for the firm to solicit the support of the various nuclear lobbies and relevant congressional delegations, as well as to track federal legislative and regulatory actions. Support for this idea was not unanimous; some members of Carlsbad's City Board objected, arguing that it was a waste of money.[26] Most of the Task Force members on the Board, however, viewed hiring a lobbying firm as indispensable. A prominent member of the Task Force pointed to neighboring Lea County's success in attracting a private nuclear facility, URENCO's nuclear enrichment plant, after hiring a lobbying firm. He argued that Carlsbad had to be similarly equipped if it wanted to "compete" with its neighbor on Capitol Hill.[27] In October 2012, the City of Carlsbad and Eddy County co-hired the C2 group, a Washington, DC–based lobbying firm, for issues related to nuclear contracts with the City of Carlsbad and Eddy County.[28] Soon thereafter, C2 hired former US Senator Pete Domenici, New Mexico's longest-serving senator, as a lobbyist. Given Domenici's familiarity with WIPP, including his role in siting the facility, the Task Force viewed him as an essential ally.

The Task Force's efforts to modify the act soon generated opposition from concerned groups in New Mexico. Although most formal political institutions in New Mexico did not oppose Carlsbad's campaign, four New Mexican environmental and antinuclear groups did. These were the Southwest Research Information Centre (SRIC), Nuclear Watch, Citizens for Alternatives to Radioactive Waste Dumping (NM CARD), and Concerned Citizens for Nuclear Safety. These groups formed a coalition to oppose Carlsbad's nuclear undertakings, creating a new website named "NO TO WIPP."[29] Their campaign organized opposition to the shipment of "hot waste to WIPP" under the slogan, "Say No To Nuclear Waste Dumping in New Mexico."[30]

The environmental groups emphasized the importance of the Land Withdrawal Act, which (as mentioned above) needed to be revised for the Nuclear Corridor vision to materialize, as a negotiated agreement between concerned groups whose democratic foundation was now being undermined. The SRIC's Don Hancock, a political scientist who had been with the organization since 1975, pointed to the years of contentious negotiations between the US Congress, the State of New Mexico, the Department of Energy, and a diverse range of concerned groups located in New Mexico.[31] For him, the act represented an act outside of time, a law that designated governmental institutions as the almost-eternal guarantors of its terms and restrictions. In striking contrast to Hancock's view, the head of the Nuclear Task Force construed the act as an ordinary agreement reached through mediation between several actors at a given moment. In Heaton's words, the "limits" set by the act were "only regulatory."[32] The language of "permanent land withdrawal" was merely administrative language that enabled local leaders in Carlsbad, at a particular point in time, to reach an arrangement regarding WIPP; by doing so, the act had fulfilled its mission. The Task Force's argument was that the terms of that political and legal arrangement, like any other law, were reversible. They could be re-questioned, revised, and revalued by a new set of actors. The sustainability of the act's terms relied on the sustainability of political will, rather than conditions "permanently" enshrined into law.

In this way, the Task Force and their environmental opponents disagreed about which moment in the resolution of a problem of rendez-vous counted as politically legitimate under the conditions of democratic governance: could consent be reached among temporally and spatially defined collectives at a specific moment indefinitely determine the conditions under which a specific piece of the Earth would "permanently" be used? When is the call for the renewal of those terms considered legitimate? When to situate "permanence" in terms of world history, in the sense that Dipesh Chakrabarty distinguishes "world history" from "Earth history" (2018), anyway? As soon as one thinks of "permanent treaties," the many "permanent treaties" between Indian nations and the US federal government come to mind. Those had been ratified to settle Indigenous nations' right to their lands, and they had regularly been abrogated through congressional votes and/or rulings at US federal courts. Then, one cannot help but ask, when does and/or should "permanence" start—specifically, in Carlsbad and, more generally, in US history (Klein 1996; Lucero 2007; Bogener 2008)? Such questions remained invisible and out-of-sight as the Task Force continued to pursue its mission to sustain Carlsbad's nuclear economy. Nevertheless, this made Carlsbad into a microcosm that helps to illuminate both the very situated and the broader problems of rendez-vous.

Keeping the Salado Formation Attached to Nuclear Waste Futures

In the spring of 2012, Carlsbad's Nuclear Task Force organized a tour of WIPP's underground facility as part of the National Nuclear Fuel Cycle Summit to showcase Carlsbad and WIPP, after the Blue Ribbon Commission published its final recommendations. For most visitors, the highlight was neither seeing disposed of waste canisters from a distance nor witnessing the trams crossing long tunnels. Rather, it was the moment they were told they could take some pieces of salt rock from the site away with them.[33] Visitors were delighted. They would be coming above ground with pieces of 250-million-year-old salt rocks, extracted from a geological formation that had been transformed into an infrastructure to contain nuclear waste.

Task Force members often referred to this same salt as Carlsbad's "most reliable asset." Sometimes they even personified it:

> I think the smartest thing we did from the very beginning was we wanted this project to be safe. And if I had to give one word or one person credit for our success, I would say it has to be the salt.[34]

The mayor almost always invoked this triangle of connecting elements—WIPP, salt, and the city—in his pitches for Carlsbad as a "solution" for nuclear waste storage. Like so many other prominent figures in Carlsbad, the mayor and his family had been involved in the potash industry for decades, and he asserted that he "knew salt and its properties."[35]

Salt was a prominent actor in Carlsbad, given that a geological repository can only operate if the geological formation in which it is located has

Figure 5.3
Pieces of salt rock extracted from WIPP's underground distributed by the Carlsbad Department of Development at WM2012.

been demonstrated to safely and securely contain nuclear waste. The Salado Formation therefore had the potential to either enable or hinder the Task Force's nuclear endeavors. During our guided tour to WIPP's underground, our guide proudly asserted that once waste is deposited at the site, the "salt would start working." He went on to explain how salt would slowly creep over the waste canisters, covering and plasticizing them, and helping them heal their own fractures. Salt and waste would become one over time, making the two very difficult to disentangle; a specific relation that has only been demonstrated between transuranic waste and salt so far.

Task Force members construed this geological property as asset: the waste's incorporation into the salt effectively locked it in place, preventing radioactive waste from entering either the water table or the food chain. However, this very property also posed an obstacle for conceiving of Carlsbad as a solution to the management of spent nuclear fuel: federal regulations require that this category of waste remain retrievable for a specified duration,[36] which led Task Force members to articulate salt's other "qualities" that could make it desirable and valuable for the containing different types of waste in the depths of Salado Formation (Méadel and Rabeharisoa 1999; Callon et al. 2002).

In his many public appearances, Heaton actively questioned the political, technical, and economic value of retrievable disposal: "Why spend billions burying only to retrieve for reprocessing?"[37] Instead, he suggested that spent nuclear fuel, which is a particularly high-heat generating waste,[38] be stored at an interim storage facility near Carlsbad, either to cool down or be reprocessed, before disposing of it in salt. Following this strategy, the government could avoid the high costs associated with retrievable disposal of commercial spent nuclear fuel.[39] To him, it was obvious that salt constituted a "better medium"; it was relatively easier and, therefore, cheaper to mine. He illustrated this viewpoint with a cost table drawn up by two local scientists.[40] With its estimates of the cost of nuclear waste disposal in different geological mediums, the table invited its reader to conclude that salt is "economically the most efficient" "solution" for high-level waste,[41] prioritizing economic efficiency as salt's primary value.

The Task Force engaged in geo-politics to align Carlsbad's subsurface with the Blue Ribbon Commission's other recommendations, including those related to the diversification of geological mediums and geological sites. For over two decades, a significant number of scientific studies had been conducted at Yucca Mountain, which features a type of volcanic rock known as tuff. Yucca Mountain is the sole tuff site in the United States; as such, the exclusive scientific focus on tuff's containment capacities had rendered the pursuit of alternative geological formations, and therefore of alternative sites, extremely difficult. The Commission advised the DOE to initiate a "generic" repository research program in various geological mediums to prevent a given research site from transforming into a de facto repository site. In response to this initiative, Task Force members created another stylized map to convey the political desirability of their geological medium (figure 5.4).

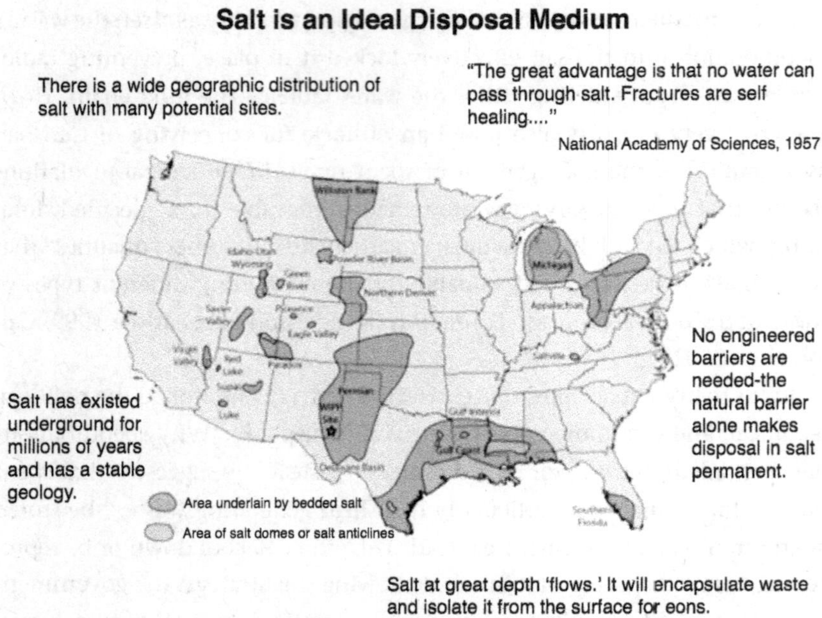

Figure 5.4
Geographical locations of salt-based geological formations.

The map illustrated the geographical distribution of salt formations across the United States, offering several possibilities.[42] In contrast to tuff, conducting research on salt formations opened the possibility of disposing of waste at other locations, not only where the research was initially conducted. In so doing, the Task Force contended that salt "democratized" nuclear waste disposal, creating the possibility to disassociate a specific research site from a potential disposal site, while opening up sites located in different geographies for consideration.[43]

The Task Force also proactively engaged in renewing governmental and intergovernmental institutions' interest in the scientific examination of salt formations. A few weeks after the Task Force presented its stylized map to curious audiences in Phoenix, Abe, my key interlocutor, invited me to attend an OECD meeting taking place via Skype at 3 A.M. Mountain Time. They, too, would see the map (figure 5.4). Abe told me that he had formed a *Salt Club*, an international group of salt experts, under the auspices of the OECD's Nuclear Energy Agency, in 2011.[44] The *Salt Club* included members from "the nations who consider rock salt as a candidate medium" for the geological disposal of high-level waste. The Club's professed goal was to stimulate "interest in other nations with appreciable rock salt deposits to consider rock salt as a viable repository medium."[45] The Club assembled nuclear waste experts from Poland, the Netherlands, Germany, and the United States (mainly Carlsbad scientists), shepherding the United States into the community of countries that reevaluated salt for high-level waste disposal, even without an official new repository science program in place.

This constituted an interesting twist for salt formations, since salt was the first geological formation that had ever been considered a potential medium for geological repository projects in the United States (National Research Council 1957). In the 1970s and 1980s, federal agencies had considered disposing of "hot" waste at WIPP. Studies were also conducted deep inside the Salado Formation and reached controversial results about salt's capacity to confine "hot waste." As the Task Force intensified its efforts to bring more waste to the site, in New Mexico, antinuclear and environmental groups started mobilizing those existing results.[46] Among others, the Southwestern Research Information Center expressed its concerns that heat generating waste may become deformed, creating instability in the geological formation and thus making salt fail to confine radioactivity.[47] My interlocutors in Carlsbad remained stoic about the potentially hostile relation between "hot" waste and salt and instead blamed the repository design based on which previous studies had been conducted. In the early studies, waste packages were to be disposed of in boreholes inside disposal rooms. Under that design, they argued, when salt would meet with the simulated heat of waste packages, it would transform into liquid, making waste canisters act like "pickles" in a bottle, resulting in a hostile environment for containment.[48] Task Force members constantly asserted the Salado Formation's suitability for containing all types of nuclear waste, but scientifically demonstrating its suitability and desirability as a medium to contain high-level wastes remained a challenging task requiring significant resources.

Reassembling Science, Salt, and the Task Force

The Salado Formation could only become a "solution for the Nation" if the new federal research program included WIPP's underground and if its results demonstrated salt's technical and scientific capacity to safely and securely isolate heat-generating wastes (whether high-level defense waste or spent nuclear fuel) from humans and the environment. Regardless of their professional backgrounds as scientists, politicians, or businesspeople, all of the Task Force members wanted the DOE to initiate repository science studies in Carlsbad.

However, the Task Force had to engage specific nuclear communities and geographies of "hot waste" and enlist their support for the Salado Formation as "the solution" for the accelerated remediation of their lands. Until the termination of the Yucca Mountain project, the Department of Energy planned to dispose of defense and commercial high-level wastes together, but the lack of an operational geological repository reopened this policy to debate, especially during the Blue Ribbon Commission's public meetings. Several DOE sites had been storing defense high-level waste in surface facilities with no end in sight, despite these having been originally designated as temporary storage sites. Key political and institutional actors at the three major sites that store this defense high-level waste (Idaho National Laboratory, the Savannah River Site, and the Hanford Reservation) began questioning whether it was necessary or even legitimate to continue with plans to dispose of spent nuclear fuel and defense high-level waste together in a single repository. They articulated several conceptual differences between commercial and defense waste[49] in terms of their lifetimes, heat levels, and levels of radioactivity; their storage locations (DOE sites versus commercial reactor sites); method of financing (taxpayers/ratepayers);[50] and their potential for reuse (high-level defense waste cannot be reused to produce energy, but spent nuclear fuel can be reused). These actors argued that the task of disposing of high-level defense waste would be made significantly less challenging if handled separately from commercial spent nuclear fuel.

In response to requests made by the political and institutional actors from the three main high-level defense waste sites, the DOE conducted a study on the heat levels of different types of nuclear waste. The study found that the decay heat of transuranic waste, such as that disposed of at WIPP, ranged from 30 to 300 watts per container, while the decay heat of a significant number of high-level defense waste canisters ranged from less than 50 to 500 watts per assembly (albeit with significant variation). In contrast, spent nuclear fuel produced between 3,100 and 4,400 watts per assembly. Officials at the sites that stored high-level defense waste interpreted these findings as an argument against their disposal at a common site, but the Carlsbad Task Force members saw an opportunity in them. The narrower the gap between the decay heat of the waste to be disposed of and that of the already disposed of waste, the easier it would be for the Task Force to assert that WIPP's geology held the capacity to contain a given stream of "hot waste," thereby reducing potential political and technoscientific controversies.

For any type of high-level waste (whether defense or commercial) to be disposed of in the Salado Formation required the Task Force to stay in alignment with the DOE's scientific and political agenda, and with each other, around several issues, including how and which interactions between nuclear waste and salt should be studied, their timing and methodology, and the institutions responsible for conducting them. The competition between technoscientists associated with New Mexico's two National Laboratories (Los Alamos and Sandia) put the alignment of the Task Force to the test. As DOE contractors, the two National Laboratories were often required to submit competing bids for federal funding. Both of their field offices wanted to ensure that WIPP's underground would be integrated into the new federal repository study, and each hoped to be the lead local agency to conduct that study. The local technoscientists therefore produced competing narratives that emphasized their institutional histories and technical expertise. They recalled Sandia National Laboratory's origins as the Z Division, an engineering and field test site of the Manhattan Project's Los Alamos Laboratory. In 1948, the Z Division—based 100 miles south of Los Alamos—became known as Sandia Laboratory. While the University of California operated Los Alamos, Sandia was operated first by Western Electric, a subsidiary of the AT&T Corporation, before coming under the management of the eponymous Sandia Corporation in 1949. From 1993 to 2017, ownership of the Sandia Corporation passed to another defense contractor, Lockheed Martin. The Los Alamos technoscientists contrasted the research orientation of their institution with Sandia's history of "corporate management," while the Sandia technoscientists extolled their engineering orientation, pointing out that their institution was the one that "acts."[51]

The local technoscientists also attempted to interest and enlist actors inside and outside of the Task Force in their cause. These efforts produced additional tensions within the larger group, as became obvious during a Task Force meeting that followed the *National Nuclear Fuel Cycle Summit* in Carlsbad. These tensions were testament to how difficult it was for Task Force members to remain in alignment with one another. The head of the Task Force called a meeting after a high-level official from the State of New Mexico complained to him about receiving "widely conflicting stories and recommendations"[52] about what kind of research project the State of New

Mexico should support at WIPP. Recognizing the danger of a situation in which each of the National Laboratories advanced its own case, Heaton attempted to rally the Task Force around common objectives.

The specific disagreements involved conflicting scientific directions for the proposed generic repository studies. Different sections of the DOE were considering different kinds of studies on salt, a different distribution of labor between the two National Laboratories, and different methods of coordinating the work within WIPP's existing site. The DOE has two major sections responsible for different kinds of nuclear waste: the Office of Environmental Management (EM), responsible for all defense waste, and the Office of Nuclear Energy (NE), responsible for all commercial spent nuclear fuel. Local actors called those two offices by their acronyms. When the DOE abandoned the Yucca Mountain project, the DOE closed the Office of Civilian Radioactive Waste Management and replaced it with a "Used Fuel Disposition Campaign (UFDC)" office. Even though the UFDC was tasked with finding a storage solution for both defense and commercial high-level waste, that is, a repository for both types of waste, it was to be housed under the NE.

In March 2011, the NE announced that it would be initiating a generic repository study in salt at WIPP: the Salt Disposal Investigation (SDI). The initial design of the study had been developed more than a decade earlier in fulfillment of the Nuclear Waste Policy Act,[53] which tasked the DOE with assessing whether it would be necessary to site a second repository by 2008. As a part of that process, the NE had funded a scoping study for the feasibility of a salt repository. The SDI was designed as an extension of that previous study. The study called for scientists to explore how salt interacts with nuclear waste at the high heat levels associated with spent nuclear fuel, specifically by separating waste packages into alcoves, which involves the creation of recessed sections inside the disposal rooms.[54]

The DOE's Carlsbad field office operated under the authority of the EM, not the NE. In June 2012, a year after the announcement of the NE's new proposed SDI study, the DOE's Carlsbad field office released its own plans for a federal repository research project to be conducted at WIPP's underground facility.[55] The plan, prepared by the Carlsbad field office, justified the new study by making reference to the recommendations of the Blue Ribbon Commission, which had called on the DOE to immediately review the implications of separating the disposal of defense high-level waste from commercial high-level waste, both of which were to be disposed of at Yucca Mountain.

The EM's proposed study focused on interactions between nuclear waste and salt at the lower levels of heat associated with defense high-level waste. This study involved placing the waste on the floor of the disposal room, rather than in the alcove.[56] This new proposal was named the Salt Defense Disposal Investigation (SDDI). The SDDI's design implied that researchers at WIPP would need to conduct in-situ heat tests that simulated the heat levels associated with high-level defense waste and use sensors to detect the heat's effect on the salt. A local DOE technoscientist contrasted the SDI's design (alcove, high heat) with that of the SDDI (in drift, low heat):

> The design continues to be re-evaluated, and since the defense waste is generally significantly cooler than previously assumed, it is going to be proposed soon that the alcove-design of the disposal system, conceptually proposed in 2008, is now seen to be too complicated, and hence expensive. With cooler waste, there is no need for alcoves to separate the waste packages widely to meet thermal goals. They can be laid in rather close proximity on the drift floor [room floor] and covered with salt. There is a lot of support for the idea that these tests are required for any sort of licensing or permitting, and we hold the opinion that to delay these tests would result in a delay to the start of the regulatory process that has to be completed before the waste is emplaced.[57]

The DOE had translated two different repository designs, targeting two different waste streams, into two research projects, both of which were expected to be completed at WIPP's underground facility. The labor and space for these projects would have to be coordinated with WIPP's Underground Research Laboratory and one or the other National Laboratory. Sandia National Laboratory primarily relied on NE funding, while Los Alamos National Laboratory relied on EM funding. The EM and the DOE's Carlsbad field office decided to support Los Alamos's field office for the SDDI study,[58] while Sandia National Laboratory expected to receive funding from the NE for the SDI.

At the Task Force meeting that was called to restore alignment by the Task Force's senior leadership, the technoscientists from the two laboratories disagreed on both the timing and the expected results of the two studies. Sandia's technoscientists argued that the two projects should be conducted sequentially, implying that Sandia would go first (with SDI)[59] and that Los Alamos National Laboratory (LANL) would proceed with SDDI only once SDI had been completed.[60] In response, LANL's technoscientists made an urgent case for their study, arguing that no high-level defense waste could be brought to Carlsbad until their study had concluded. They judged the type of defense waste associated with the SDDI as more in line with WIPP's existing geological, technical, and political properties. Moreover, they questioned whether the SDI would be able to reach positive conclusions about salt's interaction with spent nuclear fuel at high heat levels, ultimately suggesting that SDI's proponents were more interested in securing their own future than Carlsbad's. The heated debate made clear to everyone that the two institutions would have to find a way to reconcile their differences to maintain the support of local politicians and businesspeople. After a long discussion, the two laboratories agreed to pursue the studies in parallel.

There remained one further outstanding issue. Several members of the Task Force had questions about the "rumored" scientific time frames of both studies, alleged to be decades.[61] These members, mostly not scientific or technical professionals, asked whether the scientific research process could be accelerated to enable the temporality of science to be aligned with the temporality of politics. In other words, they were worried about the risk of leaving Carlsbad's fate in the hands of newly elected politicians who might oppose Carlsbad's nuclear endeavors. This time, the local technoscientists had to convince local politicians and businesspeople about the necessity of respecting the temporality of science. They did this by highlighting that scientific studies conducted at WIPP's underground would be essential for determining whether Carlsbad's underground might host other types of nuclear waste one day.

The Task Force's continued alignment on common objectives was both as essential and difficult for its members as its efforts to enlist other actors in the idea of Carlsbad as a "solution" to the problem of nuclear waste were. I saw Task Force members disputing, reconciling, and realigning their priorities, all in the attempt to maintain the "Task Force" as the unique spokesperson for all the Task Force members and to keep their interests aligned. Members' respective priorities reflected their own conflicting and complementary interests. Each member brought hybridized identities to the table, identifying as some combination of "residents," "technoscientists," and "federal officials."[62]

Conclusion

What would happen once WIPP's disposal activities concluded and Carlsbad could no longer rely on the nuclear economy? That's the question that local actors were pondering. The uncertainty surrounding the potential end of one of the major sources of their livelihood constituted the problem of rendez-vous to be resolved. By keeping Carlsbad "nuclear," the Task Force expected to retain a certain type of sovereignty and authority over its community, futures, at least temporarily and partially. Its members construed the disposal of nuclear waste in the depths of the Salado Formation, a method of making discarded materials physically "absent" from the Earth's surface (Hetherington 2004), as the bridge that ought to keep their geology, geography, and community "present" in the country's nuclear futures.

Task Force members took various forms of actions to align the resolution of their problem of rendez-vous, that of remaining valuable and desirable on the Earth's surface, with US Congress's resolution of the problem of rendez-vous, that of protecting "present and future generations" from the environmental and health consequences of radioactive matter through its containment deep inside a geological formation. This relied on attempts to hold together specific nuclear communities and geographies, the Salado Formation, specific streams of waste, and political, legal, and regulatory requirements, so that Carlsbad is characterized as "the" place for the country's nuclear waste futures. To that end, in part, they relied on proactive valuation undertakings. They created and used stylized maps, books, newsletters and coins, networks of scientific expertise, calculations, and scientific studies: all of which were to reveal the "unique" value of Carlsbad's geology, geography, and community for hosting more nuclear waste.

But what does this story tell us about the kinds of political and moral reasoning that have flourished in and through Carlsbad? Once extending Carlsbad's nuclear economy becomes the driving force of a political community, in what other ways do some of its members try to resolve their problem of rendez-vous, beyond aligning their resolution with that initially imagined by the US Congress, and beyond attempts to bridge human time and geological time?

6 Entrepreneurial Displacements: Yes in Our Backyard and the Reimagining of Nuclear Waste Futures

In the previous chapter, Task Force members pursued a resolution to their problem of rendez-vous that relied on reconceptualizing the City of Carlsbad and WIPP together as an ideal repository site for the long-term disposal of nuclear waste. Numerous local actors depicted the city and its extant plant as intermediaries on the Earth's surface that would facilitate the transfer of nuclear waste to deep inside the Earth's subsurface, thought to ensure the protection of very broadly temporally and spatially defined collectives from the undesirable consequences of radioactivity while preempting the transformation of their community into a remain of the nuclear age. In this chapter, I analyze how the nuclear waste futures posited for Carlsbad represent key transformations in the political and moral reasoning that had once guided the formulation of the national nuclear waste policy.

Extending the inquiry into the resolutions of problems of rendez-vous, I unfold how Task Force members, looking to maintain Carlsbad's nuclearity, create a particular kind of political community, one composed of entrepreneurial citizens who turn what had previously been conceived of as a national responsibility into an opportunity for local enterprise. As historian Benedict Anderson highlights, a political community is both imagined and invented where it does not previously exist (2006, 6). Anderson does not use the notion of invention to suggest that political communities are simply fictional but rather invites us to pay attention to the styles that people engage in imagining, inventing, and creating those political communities. As studies of sociotechnical imaginaries have emphasized, imaginaries not only are about the past and present but can also be "futuristic and instrumental" (Jasanoff and Kim 2009, 2015), as is apparent in the case of Carlsbad.

This chapter starts by examining the political community imagined by Task Force members before unfolding how some of its members engaged in positing an alternative resolution to Carlsbad's problem of rendez-vous: one that involves the containerization of nuclear waste. With this resolution, nuclear waste can be transmuted into mobile objects and then transported to and "temporarily" stored in the region of Carlsbad, thereby becoming an object of new forms of economic transactions. The entrepreneurial imaginary and its related strategies of containerization represent more than a new chapter in the material logistics of dealing with radioactive matter. Choosing the City of Carlsbad as an emplacement for nuclear remains, perhaps counterintuitively, enables a myriad of displacements of risks and responsibilities across time and space.

Eternal Wilderness Areas, Eternal Wastelands

As soon as I arrived at Carlsbad, I was invited to take a guided tour of the famous Carlsbad Caverns, one of the largest limestone caves in the world, located in a fossil reef laid down by an inland sea 250 to 280 million years ago. Abe, my key interlocutor, visited the Caverns almost every weekend, whether alone or with family, friends, visitors, or colleagues. He had first discovered them following an invitation by colleagues who were members of the local caving club to do some restoration work at the Lower Cave. "Never have I seen such rich abundances of stalagmites, stalactites, and blades of calcite hanging from ceilings, walls, and reaching up from floors. It is unimaginable, it needs to be seen to be believed," he reminisced about his first impression.[1] Humbled by nature, Abe often invited me to come on other hikes in the region. During those hikes, he always enjoyed taking breaks to silently contemplate the wilderness surrounding us—the very landscapes many construed as a solution to the problem of managing nuclear waste.

Several scholars have examined how wilderness areas and wastelands become intertwined, perhaps eternally (Galison and Moss 2015; Houdart 2018). In Carlsbad, where this entangled relation is part of the residents' daily lives, it allows for legitimate claims about the "unique value" of Carlsbad as a place where wilderness and radioactivity have peacefully coexisted. This narrative simultaneously tells a story about the past and provides an interpretation of Carlsbad's present state in ways that facilitate local actors' attempts to secure a nuclear future for themselves and their city.

Toward the end of my fieldwork, the head of the Task Force offered me an expressive book titled *Carlsbad* (Marshall 1998b), a tome recommended in local reviews as the only existing book on the history of the city and its immediate surroundings. The book's author, Terry Marshall, was both a local resident and a rural sociologist. He was also someone who had closely studied Carlsbad's waste management industries. In addition to his book on the city, he had conducted a study before operations began at WIPP. The resulting report, titled "Socio-economic Impact Study of the Waste Isolation Pilot Plant," was published during the same year his book was released (Marshall 1998a). Released in limited edition, the book was given to external visitors like me, suggesting an endorsement of its portrayal of Carlsbad. This resonated with ideas that Task Force members frequently enunciated, not only during our discussions and interviews but also during regional walks and hikes they invited me to take with them.

Several things stood out to me about *Carlsbad*. First is the speed with which it traverses time periods and spatial sites. Within a few pages, the text takes the reader on a journey through the history of the city from the "Old West" to a "retirement city," from a "geological repository" to the site of Carlsbad Caverns and the national park located nearby. Neither the Old West nor nuclear waste disposal is presented as any more "local" than the other. The book constantly accentuates the cohabitation of tradition and modernity, local and national, and nature and built community, whether moving from the creation of the city to the region, from the creation of transportation routes to the discovery of Carlsbad Caverns and its transformation into a national park, or from the potash mining industry to the siting of a nuclear waste repository. Marshall defines Carlsbad as a "desert oasis," "a geological paradise," with a unique nature of "colors and a harsh beauty" and "a vast underworld." Temporally, the book underscores the perpetual nature of the landscape, in both its geologically and social dimensions:

> To Carlsbad, the Carlsbad Caverns National Park is far more than a natural wonder set aside for national use. It is our community's world-class heritage. Park personnel change. Park policies change. But the Park remains ever vital to our community. We will always remain active in promoting and protecting our Caverns. (Marshall 1998b, 23)

Marshall covers the city's history in a way that produces a past, present, and future that is continuous, entangled, coherent, and consistent. In the Introduction, he presents this story in a way that discards the history of the Indigenous peoples who had been living on that land for thousands of years before settlers arrived:

> By 1884, New Yorker Charles B. Eddy was grazing cattle where Carlsbad now sits. Then, as happened so often in our history, nature set a new course. Disastrous drought between 1884 and 1886 wiped out a third of Eddy's cattle. He had to find a more stable income. Eddy and Pat Garrett (of Billy the Kid fame) incorporated the Pecos Valley Land and Ditch Co. to irrigate land.
>
> A town named Eddy followed, formed on the south bank of the Pecos late in 1888. Eddy and his partners set a tone of unabashed boosterism that survives to this day. They started a newspaper, hired a photographer, and printed enticing brochures to lure settlers to a new land of milk and honey. They built a land from scratch in the treeless desert. They undertook a grandiose water dam project. Nature retaliated- floods in 1893 and 1904 nearly destroyed the town, and the Pecos flooded more than a dozen times over the years. Winter freezes wiped out the area's flourishing orchards. Despite setbacks, boosterism and the tireless pursuit of grand dreams lived on. In 1907, city leaders convinced the U.S. Department of Interior to rebuild the town's irrigation system. A year later, they led the push to build a road to El Paso. They built an airport in 1926. They promoted Carlsbad Caverns, succeeding in having them named a national park in 1930, and getting a paved road built in 1934. In the 1940s, Carlsbad soldiers won fame on Bataan; at home the Carlsbad Army Air Field trained airmen for the war in the skies. (Marshall 1998b, 8)

The author continues by presenting the city's major economic activities—potash mining, oil drilling, and nuclear waste disposal—these activities of disposal and extraction exist on a continuum. There is no rupture or inconsistency generated by the arrival of nuclear waste. On the contrary, nuclear waste disposal is presented as a natural extension of potash mining. Quite similar photographs are used to depict potash mines and the geological repository for nuclear waste. The geological repository is distinguished only as a more orderly and technically sophisticated facility.

In accordance with Marshall's viewpoint, Task Force members tended to link mining and nuclear waste disposal, often doing so by referencing the concept of "safety culture." At that time, they explained the "safety record of WIPP" with reference to a preexisting safety culture that had developed in Carlsbad's mining community, a culture they argued had been necessarily enhanced with the arrival of WIPP and its codified working procedures, practices, and standards. In his interview with *Forbes* magazine, the mayor contrasted his work as the safety director at a local mine to the work at WIPP: "All told, Carlsbad's potash mines hold more than 1,000 miles of tunnels covering 100 square miles. Compared with going after potash, says Janway (the Mayor), digging a spacious mine to hold drums of waste is easy. The miners were all for it" (Helman 2012). In Carlsbad, the geological disposal of nuclear waste and potash mining were construed as intrinsically interwoven activities that produced the successful performance of WIPP. Depicting a deeply rooted safety culture in this way set the stage for expectations that nuclear waste management in Carlsbad would only continue and enhance this record.

"YIMBY" Success in a "NIMBY Nation"

Let's return to the *Carlsbad* book. A reader encounters another slogan that was often repeated by the Task Force members: Carlsbad was "a 'we-can-do-it' town" (Marshall 1998a, 61). The author depicts people in Carlsbad as members of an entrepreneur community that constantly determines its own destiny while remaining loyal to its traditions and its past. This depiction is not limited to the siting of WIPP, but an extension of the Carlsbad community's general philosophy of life. The city's history is further described as a battle for survival. Carlsbad's mastery of its own destiny involves its community's mastery and control of nature. It is a city built from "scratch" where nature is discovered, isolated, and protected, and nuclear waste is simultaneously contained in ways that protect the natural and built environs.[2]

Task Force members drew upon a similar notion of an entrepreneur community. They, for instance, construed the disposal of nuclear waste in the depths of the Salado Formation as an outcome of their search for an alternative to the extraction of mineral resources from the same geological formation. Until its decline in the 1960s, Carlsbad was mainly a potash mining community, and the Salado Formation had been and continues to be a major source of potash, mined from bedded deposits of sylvite and langbeinite.

In the United States, the exploration of domestic supplies of potash began in the 1920s following Germany's embargo on potash exports in 1915, and it accelerated with the increased use of potash in the production of agricultural fertilizers. In 1925, potash was discovered in Eddy County. By 1944, New Mexico surpassed all other domestic potash producers, providing 85 percent of the total national production of 5.7 million tons (McCutcheon 2002, 21). However, unable to compete with the continued penetration of foreign producers into US domestic markets, especially following the discovery of potash reserves in Canada, by the late 1960s, Carlsbad's economy was devastated by the shutdown of the major potash mining companies, including the US Borax and Chemical Company. This decline led to more than a thousand jobs being wiped out.

Task Force members often started our conversations about nuclear waste disposal in Carlsbad by narrating how the nuclear economy had developed as a result of local initiatives related to the collapse of the potash mining industry. They situated the local community at the origin of WIPP and elided an account of themselves as a community that was designated by the federal government to host a repository project, or upon whom such a project had been imposed.[3] They noted with pride that local political leaders had successfully negotiated the installation of local offices for the two National Laboratories and the Department of Energy (DOE) for activities related to nuclear waste disposal. One Task Force member explained that no such offices were in place in the 1970s when studies began. Employees arrived on Monday mornings by plane from Albuquerque, the capital of New Mexico, and returned on Friday evenings.

"Boosterism," "we-can-do-it," and "being a solution" are phrases and terms that Task Force members invariably mobilized to identify the Carlsbad community in these same conversations, before asserting that the federal government had "accepted" WIPP after the community "itself" had negotiated the terms and conditions of the project. Their characterization of Carlsbad as an exemplary entrepreneurial community has received attention beyond the city's borders. The lengthy article about Carlsbad in *Forbes*, whose publication coincided with the end of the Blue Ribbon Commission's mission, marked a moment when Carlsbad gained high visibility at a national level (Helman 2012). The Carlsbad community was depicted by the *Forbes* journalist as an example for the nation and as a significant manifestation of what he referred to as the emerging "Yes in My Backyard" (YIMBY) nation. Displaying the opposite ethos of "Not in My Backyard (NIMBY)," an acronym that refers to local communities that oppose the siting of infrastructures due to different types of disturbances they are expected to cause in their everyday lives, YIMBY designates local communities that say yes to hosting such developments. Marshall puts it,

> More importantly, Carlsbad retains its can-do-it attitude, a sense that our future is up to us. We remain optimistic in promoting our city with almost the same shameless optimism and boosterism that marked the town at its very inception. (Marshall 1998b, 9)

According to the *Forbes* article, the difference between these two groups was not limited to refusing or agreeing to host nuclear projects. While

NIMBY communities sought growth driven by the "financial industry" and "asset inflation," YIMBY communities like Carlsbad hosted "indestructible and inescapable industries" producing real "growth" in "hard pressed areas," and thus were doing "the country's dirty work":

> "When it comes to doing the country's dirty work, Carlsbad is hardly alone. Hard-pressed areas across the country are coming to the grown-up conclusion that they will never become hotbeds for venture capital or attract Stanford or Google (or even Toyota) to open a satellite campus. Instead, they're inviting indestructible and inescapable industries, like prisons, dumps and oil and gas drilling to town—fostering Yes In My Backyard success (YIMB) in a Nimby nation. There's big parts of the country that don't want to get their fingers dirty," says FORBES contributor Joel Kotkin, the renowned demographer and author of The Next Hundred Million: America in 2050 (Penguin Press 2010). In prime Nimby states like New York and California, says Kotkin, "They don't want growth; they want asset inflation." (Helman 2012)

Such an exemplary community accepted they were excluded from certain futures, had limited expectations about what their futures held, and took responsibility for their own destiny. Composed of "rational actors," they acknowledged their condition and did not expect a welfare state to care for them. The way the Carlsbad community is characterized by the journalist as an exemplary, independent, responsible, and entrepreneurial community resonates not only with traditional ideals of white settler masculinity but also more specifically with what Emilie Hache (2007a, 2007b) calls the "neoliberal pastorate." Just like pastoral Christianity, which is considered to govern all features of life, the "neoliberal pastorate" removes distinctions between the political spheres and the economic sphere, between the private sphere and the public sphere. Under the "neoliberal pastorate," as distinct from pastoral Christianity (Hache 2007b), neither obedience nor submission to a higher or an external power is praised. The State with a big "S" does not directly intervene. Rather, traditional themes of responsibility are transferred to private institutions and individuals. This requires individuals to transform certain undesirable behavior or forms of action, necessitating their detachment from their dependence on the State by simultaneously rendering a different form of conduct desirable: one that is qualified as "responsible." The ultimate step to exemplarity is the maximization of self-will and not the elimination of the self—as is required in the pre-neoliberal Christian pastorate.

Whereas environmental and antinuclear groups located in northern New Mexico frequently denounce the Cold War nuclear project for having produced "environmental ruins," "geographies of sacrifice," and "wastelands" (Kuletz 1998), local actors reflected on the modernization and economic growth that the State of New Mexico and other nuclear geographies experienced. On the one hand, the Task Force members associated Carlsbad with these nuclear geographies and aimed to be one among them, while on the other hand, they distinguished themselves from those who contested the continued presence of nuclear facilities and their radioactive remains within their communities, geologies, and geographies. To do this, they engaged in rhetorical practices of "othering" opponents. For instance, Reverend Rogers, a local pastor in Carlsbad, sometimes gave his Sunday sermon at the First Baptist Church on fear and how "fear is an enemy to be beaten."[4] He was a member of the Task Force. When the Blue Ribbon Commission visited Carlsbad and held its public hearing, Reverend Rogers was one of the local residents who gave testimony, during which he denounced nuclear fears as "false appearing real," and underlined how Carlsbad is beyond that fear:

> I remember the memory of my deceased grandfather, Richard Rogers, a veteran of World War Two, who said over 60 years ago that his only regret with nuclear anything was that it was introduced to the world through atomic weaponry. It was his contention then that that would forever taint the genuine potential, promise; the possibilities and the great importance that nuclear would have for the future our planet. We have had years of incredible propaganda, and a campaign fuelled on Cold War fears. We have had lots of fearmongering postulated by questionable theatrics and emotionalized rhetoric to stimulate and to motivate the fear. Here in Carlsbad, we have no fears. As a citizen of this community, I am here to say that fear does not govern us. It does not determine how we choose to interpret the empirical data. . . . Remember what I have said in the past on other things, fear is false evidence appearing real. When you evaluate this process, keep in mind the thoughts of those of us who live with WIPP every day, and we are not afraid. We are ready to embrace the future because we believe in WIPP.

Without specifically identifying those who "fear," Reverend Rogers, like other Task Force members, mediated the reappraisal of the conduct of the Carlsbad community as "exceptional" and "exemplary" and suggested their superiority. They engaged in acts of valuation: not only by projecting the community as desirable and worth achieving but also by situating "self" as hierarchically superior vis-à-vis the "other." Further, and importantly, they did this within an increasingly polarized political terrain in which their characterizations of Carlsbad as independent, self-reliant, and willing to do the country's dirty work echoed free-market conservative rhetoric.

This identification of the Carlsbad community in opposition to NIMBYism has extended from other connotations of the label. Risk scholars who conduct risk assessments have historically used NIMBYism to negatively denote local opposition movements[5] against the siting of nuclear waste facilities in their vicinities, which the locals deem to be risky or simply undesirable.[6] The authors of these studies have argued that opposition movements are the result of an inefficient distribution of "externalities." They point out that the monetary and psychological costs of a hazardous facility were incurred by people living in its proximity, while its benefits are distributed throughout the entire economy. Based on that approach, they have suggested that higher "acceptance" levels could be reached through monetary compensations that would mediate the efficient redistribution of "externalities."[7]

To determine the level and type of compensation, the authors of such studies have relied on techniques borrowed from economics, such as contingent valuation and auction methods. Their explanation that the opposition movements emerged due to inefficient resource allocation was situated at the heart of various socioeconomic and environmental assessments employed by the federal government. It also resonated with industrial actors who identified members of these groups as egotistic and self-interested, and either resistant to sacrificing their own self-well-being or ignorant. American historian Gerald Jacob, for instance, has pointed out that authors of environmental impact assessments concerning a repository siting construe nuclear waste disposal as a "locational problem" (Jacob 1990, 157). He notes how such assessments reduce opposition movements to "locational conflicts" and limit a range of externalities to an estimation of "how many fire hydrants, tennis courts, and new apartments would be needed and at what price." This results in the definition of a wide range of local concerns as identifiable, quantifiable, and solvable, and in the notion that these concerns could be "calculated, bargained over, and negotiated to settlement" (Jacob 1990, 158).

The Task Force members never accepted the identification of their community as one that tries solely to maximize its utility. One example of this is that when the first layoffs took place at WIPP, they tried to trade off the annual compensation they had been receiving from the Department of Energy as an "affected community" against those layoffs, in order to keep the city "nuclear." This symbolic gesture was important to them since, to a certain extent, they thought it demonstrated that they did not simply want more nuclear waste purely for the economic benefits. They associated the pursuit of their own well-being with that of the nation. In conversations with me and in public statements, such as the following given to the Blue Ribbon Commission, they implicitly contrasted "the other's" irrationality and fear (of nuclear waste) with the way they construed the Carlsbad community: one made up of an assemblage of "citizens" devoted to "serve the nation":

> Carlsbad's citizens are one of a kind. The citizens are open-minded, caring, giving, hardworking and have a great can-do attitude. The community is willing to pull up their sleeves and help out in any situation or tackle any problem. The community wants to be a part of the solution, not a part of the problem. Carlsbad was the only community in the nation that stepped up to the plate and asked for the WIPP project here. Carlsbad took the initiative to become educated about nuclear waste before making the decision about WIPP.[8]

While this implicit and partly explicit denunciation of the "other" enabled "self" praise, this identification of the "self" was never solely symmetrical to the "other"; for these actors, the Carlsbad community had taken responsibility for the future of the nation and not solely its own future. By invoking the nation and the national interest, they endowed their nuclear aspirations with a higher moral objective and made no distinction between individual and collective action or between the common good and private interest.

During the Blue Ribbon Commission hearings, one of the Task Force members also spoke with passion about Carlsbad's "patriotism" and their will to "help the nation."[9] However, while he associated the community's commitment to hosting nuclear waste with the well-being of the nation, he did not expect the federal government to be the agent siting new nuclear projects in Carlsbad. Rather, he expected the federal government to enable Carlsbad to become the "solution" for the "nation." He wanted them to remain entrepreneurs of "their" own futures. A local technoscientist and Task Force member further spells out this entrepreneurial stance during one of our conversations, addressing any people with objections to their nuclear endeavors in the following way:

> We have something in Carlsbad that no other place in the nation can ever have. We have an operating repository. We don't have to promise that it works; we can show you that it works! It has been thirteen years that we have been burying waste. What is the problem? What is your problem? The safety puts South-eastern New Mexico to a higher place than any other place in this country.[10]

From this perspective, the Carlsbad community was composed of entrepreneurial citizens who acted for the well-being of the "nation" and not only in the interest of its own well-being. In that sense, the characterization of Carlsbad as a valuable and desirable community resonated not only with notions of neoliberal responsibility but also with neoconservative patriotism, defined as the activity of "preserving what you've got and protecting your own, whether an individual family or the national family" (Brown 2006, 701). As Wendy Brown suggests, the encounter between neoconservatism and neoliberalism often results in clashes but also gives rise to unexpected results. The liberal democratic mode of state, Brown explains, is founded "on universal, procedural and juridical" values that are "religiously and culturally secular, and as peaceful and defense oriented" (Brown 2006, 701). However, she goes on to explain that in a form of government that is situated at the articulation between neoconservatism and neoliberalism, the

> nodes of legitimacy are replaced by a figure of a state that is openly partial, maneuvering, and political; openly invested in culture and the market; openly engaged in promoting a civic religion that links family form, consumer practices, political passivity, and patriotism; and openly and aggressively imperial. Each of these reformulations is significant unto itself, but together they establish a relation of mutual reinforcement between newly legitimated statism in domestic and international politics.

Brown draws our attention to how the coexistence of neoconservative and neoliberal forms of government results in the devaluation of political autonomy, the transformation of political problems into individual ones with market solutions, and the definition of citizens as consumers who are yet available to a legitimate statism. In Carlsbad, this encounter appeared natural. Entrepreneurial citizens claimed that they had the capacity to accomplish the strength of the state and asserted that the rights and obligations, as well as the interests and desires of the community—which they claimed to represent in totality—could be achieved through their entrepreneurial actions.

From Containment to Containerization of Nuclear Waste

As entrepreneurial citizens, not all Task Force members judged that Carlsbad's nuclearity had to rely on the physical containment of nuclear waste in the depths of the Salado Formation. At the time the Yucca Mountain project was terminated, nuclear waste management relied mainly on two techniques of containment: the actual and the imagined and planned. Construed as complementary to each other, both were based on the anchoring of spent nuclear fuel in place, out of sight. In practice, spent nuclear fuel assemblies were contained "temporarily" in spent nuclear fuel pools at their place of production. In the planned "permanent" option, they were to be contained deep inside the Earth's subsurface, at a location designated by technoscientific and political institutions.

Yet, as the only existing "permanent" option was off the table with the abandonment of the Yucca Mountain project, the inventory of spent nuclear fuel was continuously growing. By 2011, existing nuclear power plants in the United States had produced approximately 65,000 metric tons of spent nuclear fuel, stored at more than seventy-five sites in thirty-three states, while 104 operating nuclear reactors generated about 2,000 metric tons of spent nuclear fuel annually (US Government Accountability Office 2011b). This resulted in the densification of assemblies stored at pools. Nuclear experts had started raising concerns about the safety and security of such storage, calling for the adoption of an alternative method of storage: dry cask storage systems. In particular, the Blue Ribbon Commission revived this idea by recommending that, until a repository becomes available, spent nuclear fuel assemblies stored "temporarily" in pools should be transferred to dry casks and stored at a reduced number of sites (i.e., "regional" and or "centralized" "interim" storage facilities). Carlsbad, the Task Force imagined, would be a natural fit to become one of these sites. Some of the Task Force members considered that the siting of a consolidated storage project around Carlsbad could constitute a more immediate and attainable way to resolve their problem of rendez-vous, that of remaining "nuclear" and, therefore, valuable on the surface, without having to transform the entire technoscientific, legal, political, moral, and regulatory order that once made Carlsbad "nuclear."

Spent nuclear fuel pools consist of steel-lined, seismically designed concrete pools that keep thermally hot and highly radioactive spent nuclear fuel assemblies under 20 feet or more of water. Water serves two purposes: to absorb heat and to shield workers from radioactivity. Additionally, water pumps supply continuously flowing water to cool the spent fuel, while water-level monitors and radiation detectors are also placed in the pool to detect when to replenish water. Laid out in the pool, fuel assemblies become visually inspectable and therefore manageable, enabling nuclear operators to act upon them in case of problems. Over time, as active materials, fuel assemblies become thermally cooler as their radioactivity decays. This opens the possibility for them to be contained in other forms (after at least five years).

Dry cask storage systems rely on surrounding spent nuclear fuel assemblies with inert gas and placing them inside containers called "casks." In a way, it is a form of containment that is made possible by the replacement of the water found in the pool, a concrete shell, with inert gas placed within another concrete shell. These casks are typically steel cylinders that are either welded or bolted closed. Each cylinder, surrounded by additional steel, concrete, or other material to provide radiation shielding, is placed vertically or horizontally in a concrete vault. Flushing out water and replacing it with a more inert substance not only raises the limit of storage set by the specific size of the pool but also leads to the transformation of spent nuclear fuel into a self-contained, mobile object. Dry casks constitute an "envelope" that places spent nuclear fuel in a state of temporal and spatial "suspension," allowing it to be stored and circulated "beyond its place of production" (Kenner et al. 2019).

How casks work is quite straightforward—they isolate heat and radiation. Yet this containerization technology inhabits the potential of a consequential transformation: not only in the political economy of nuclear waste but also in the way nuclear waste futures have been imagined. Containing spent nuclear fuel in this form enables its displacement (and risks associated with it) to other sites (that may or may not be located at a reactor site). In that sense, dry cask storage provides industrial actors with the possibility of transforming spent nuclear fuel into the object of economic transactions and capitalistic exchange.

Dry cask storage systems were not new. In the late 1970s, following the stagnation of its nuclear program and the growth of its spent nuclear fuel inventory, Germany developed these casks to increase the onsite storage capacity of its reactors. In the United States, the first dry cask storage installation was licensed in 1986 at the Surry Nuclear Power Plant in Virginia. As these installations were costly, for a long time, nuclear utilities transferred waste from pools to dry casks only when they could no longer materially and legally extend the storage capacity of their pools (Janberg and von Hippel 2009).

In fact, consolidated storage facilities were also given a specific place within the framework of the national nuclear waste policy. The Nuclear Waste Policy Act of 1987 called these "monitored retrievable storage facilities."[11]

Policymakers had approached this new mode of containment with caution. The continued storage of spent nuclear fuel at such facilities was assessed as a risk that could lead utilities to fail to use (and thus finance) federally built geological repositories. The legislation banned the Department of Energy from building, owning, and operating them unless there was an ongoing geological repository program. It also subjected them to specific design constraints. Their construction could only start once the Nuclear Regulatory Commission licensed a geological repository, and the total volume of waste stored on site was subject to size caps: 10,000 metric tons of heavy metals (MTHM) until a "permanent" geological repository started accepting waste and 15,000 MTHM thereafter. Through these design constraints, legislators who drafted and passed the act aimed to prevent these facilities from undermining the "permanent" resolution to the problem of rendez-vous, as laid out by the national policy, which relied on the containment of waste in the depths of a geological formation for thousands of years.

The passing of the act had led to the creation of the Office of the Nuclear Waste Negotiator, which was charged with finding communities willing to site this type of facility (Stewart and Stewart 2011, 236). The first "Negotiator" took office in 1990 and invited potential "host communities" to apply for a series of study grants to fund research on the effects of hosting a "monitored retrievable storage facility." Twenty-four Indian nations and four counties in four states applied to receive the grants. State governments then vetoed the applications of counties under their jurisdiction. This left Indian nations, who held sovereign authority over their tribal land, as the only candidates.

Randel D. Hanson (2001) describes that offers to host such facilities came at a specific moment for many of these Indian nations. On the one hand, the nations were experiencing an expansion of their "political sovereignty over tribal lands in an era of self-determination," which for some had been happening since the 1960s, while on the other hand, they lacked the means to allow their reservation economies to be self-sustaining. Their situation worsened when the Reagan administration argued that its federal trust responsibility toward American Indians was "a form of socialism or social welfare or both" and decided to reduce trust monies, which increasingly forced the people of the nations to "seek marketization opportunities for economic viability" (Hanson 2001). These developments had the effect of making neoliberal practices one of the few ways in which Indian nations could accomplish political and economic "freedom," and they made "voluntarism" into "the agent of incorporation" of Indian life and land into Western capitalism (Hanson 2001).

These initiatives resulted in various contestations. Environmental justice advocates opposed such projects for imposing hazardous facilities on poor communities. State legislatures opposed them for having enabled sovereign Indian nations with small populations and limited land to impose major negative externalities on their surrounding regions.[12] In 1994, following the failure of the "Negotiator" to reach an agreement among concerned actors about the siting of such a facility in a locality, Congress decided to disband the Negotiator's office (Stewart and Stewart 2011, 237). From then on, consolidated storage initiatives periodically reappeared. With the federal government failing to meet the terms of its national nuclear waste program, the private nuclear utilities decided it was time for them, as market actors, to take action. They decided to find ways to deal with their "externality" on their own terms: by building consolidated storage facilities, as private initiatives were not subject to the same design constraints as those initiated by the federal government.

In 1996, Northern States Power Company, a nuclear utility in Minnesota, initiated a project named "Private Fuel Storage." In 1992, the Minnesota state legislature had limited the company's ability to expand its spent nuclear storage capacity at its two reactors. This ruling, alongside the failure of the federal government to have an operational repository, had led the utility to start seeking other means to handle its waste (Rogers and Kingsley 2004). Eight private utilities united around the project, with a plan to own and operate a consolidated storage facility with enough space to store spent nuclear fuel from other utilities (40,000 MTHM in total).[13] These utilities picked up where the office of the Negotiator had left off in previous negotiations with Indian nations. The outcome of this new initiative was that the Skull Valley Band of Goshute Nation, located within the frontiers of the State of Utah, agreed to lease a portion of its reservation for the construction of the facility in exchange for lease revenue and employment opportunities from both permanent and temporary construction jobs.[14] The initiative, however, then failed, following opposition from the State of Utah, even though Private Fuel Storage had been licensed by the Nuclear Regulatory Commission in 2006. After the shutdown of the Yucca Mountain project, initiatives to site surface facilities to store spent nuclear fuel regained momentum in Carlsbad and in its region.

Just a few days after the National Fuel Cycle Summit, Alaska Senator Lisa Murkowski paid a visit to Carlsbad. Senator Murkowski was chair of the Senate's Energy and Natural Resources Committee and thereby held one of the key Senate positions influencing US energy policy at that time. Murkowski was also a supporter of the Yucca Mountain project. She had come to Carlsbad with Heather Wilson, the New Mexico candidate for the Senate who expected to replace the retiring Democratic Senator Jeff Bingaman. Both Wilson and Murkowski attended a fund-raising luncheon, where attendees had been required to donate to Wilson's election campaign. The Task Force members were happy about the visit. They believed that it signaled the senator's support for their nuclear endeavors. When Senator Murkowski then took a guided visit to WIPP's underground and, in an apparent last-minute decision, expressed her desire to talk to and hear from members of the Carlsbad community, members of the Task Force and the city officers sent an email to invite their contacts to give testimony. In less than two hours, a large conference room was filled by the Task Force.[15] Even though it was not an official public hearing, the conference room was configured in a similar setup to that of a public hearing. The mayor was the first to take the floor. During this "rituals of assent" (Gusterson 2000), each individual conveyed their prepared message in less than two minutes, in an accomplished performance. A few repeated the same statements they had previously delivered before the Blue Ribbon Commission.

Two officials from a major national nuclear trade association and executives from three nuclear utilities were also at the conference center, having accompanied Senator Murkowski during her trip to Carlsbad. As the senator listened to the public comments, a parallel meeting took place close by.[16] A senior manager from the national nuclear trade association and one of the association's nuclear waste experts, as well as the executives from three major private nuclear utilities (the Connecticut Yankee Atomic Power Company [CYAPC], Xcel Energy, and Constellation Energy), met in the adjoining room. Connecticut Yankee is a nuclear power plant that was fully decommissioned at the time. Together with two other companies that had decommissioned nuclear power plants since 1998 (Maine Yankee Atomic Power Company and Yankee Atomic Electric Company), they had been previously and regularly awarded monetary damages paid by the federal government.[17] At reactor sites, all that was left was stored containerized spent nuclear fuel.

Even though nuclear utilities were awarded compensation, as nuclear power plant licensees, they still had to manage those sites and remained legally liable for the safety and secure storage of spent nuclear fuel. Xcel Energy owned and operated two nuclear power plants: Monticello Nuclear Generating Plant and Prairie Island Nuclear Generating Plant. Constellation Energy, together with Electricité de France,[18] owned Calvert Cliffs, Nine Mile Point, and Ginna Nuclear Generating Facilities. A month before the meeting in Carlsbad, Constellation Energy had merged with Exelon Energy, which owned the largest number of nuclear power plants in the United States.[19] The presence of these three executives in Carlsbad with three different profiles was indicative of how consolidated storage facility projects interested a wide range of actors in the nuclear industry at this time.

In fact, the continued storage of spent nuclear fuel at reactor sites was not solely a concern for industrial actors or the federal government. The end of the lifetime of these reactors signaled the end of many local nuclear economies. In places where shutdown, decommissioned, or soon to be decommissioned nuclear power plants operated, some vocal and high-profile members of local communities were increasingly impatient to see their lands remediated and allocated to new activities, with the number of these sites only expected to increase as more reactors approached the end of their operational lifetimes. Some saw in containerization new opportunities for their communities, and they began engaging in activities in preparation for participating in such enterprises.

However, the mobility of spent nuclear fuel and its storage in places other than its site of production also generated controversies. Not everyone agreed that spent nuclear fuel should be circulated across the country and become objects of economic transaction. Reviving a Senate bill previously proposed as a reaction to Private Fuel Storage, in 2010, more than 170 national and regional environmental and antinuclear organizations in fifty states assembled against the construction of centralized storage facilities by signing a manifesto they named "Principles for Safeguarding Nuclear Waste at Reactors."[20] The manifesto contended that spent nuclear fuel should stay where it was produced and that storing spent nuclear fuel as closely as possible to its place of generation would avoid associated transportation risks. For these actors, the safe and secure storage of spent nuclear fuel until its geological disposal did not require consolidated storage, and it was not ethical to transfer associated risks to communities and states that had never benefited from nuclear energy. These voices regarded consolidated storage as a last resort, to be used only in the event of a reactor site becoming unsuitable.[21]

In Carlsbad, the idea of building a consolidated storage facility fell both within and beyond the scope of the Task Force's core mission. The siting of such a facility in the region of Carlsbad was a past future, one that local businesspeople and politicians wanted to revive. Under the George W. Bush administration, a Global Nuclear Energy Partnership (GNEP) program was launched to revive nuclear energy production through the reintroduction of reprocessing of spent nuclear fuel and the development and implementation of alternative nuclear fuel cycle technologies. This project required new sites to host related facilities. In response, Eddy and Lea Counties had formed what they named "the Alliance" and officially volunteered to jointly host GNEP facilities.[22] The two counties, which normally competed to host new nuclear investments, had taken inspiration for the idea of a cross-border alliance from the two bordering French communities that host the national underground research laboratory for the French geological repository program and share the associated economic benefits.[23]

To replicate the French model, Eddy and Lea counties had bought a piece of land located at the border of both counties. Once the Alliance had volunteered to host the GNEP facilities, they received federal funding, which they used for the preparation of an environmental impact assessment on that specific piece of land. Despite the Obama administration and Congress later abandoning GNEP, the Alliance continued to maintain both the land and their mutual engagement. As soon as the Yucca Mountain project was abandoned, they then decided to collectively lobby for the siting of a consolidated storage facility on their jointly held land, thereby reanimating that aborted future into a present one.

Leaders of the Eddy–Lea Alliance approached this meeting with leaders in the nuclear industry as a common effort to find the right "business arrangement." Parties had presented their demands and discussed several cost elements: the cost of the license application to the Nuclear Regulatory Commission, the cost of transferring the spent nuclear fuel from pools to dry casks, and the savings that the nuclear utility companies could make by no longer having to keep those materials on their individual reactor sites. The meeting had also included a discussion about the monetary incentives that the State of New Mexico and the municipalities would receive "per cask and per year" during the licensing period of the facility. Compensations were construed as a rental fee, which was a sentiment picked up by antinuclear groups, who named the proposed facilities "parking lots."

The members of the Alliance, among whom were some of the members of the Task Force, were concerned about the "failed example" of the Private Fuel Storage. What this showed them was the importance of having the right "business model." They judged that the project's failure had resulted from its designers' attempts to short-circuit the state authority and insisted that they instead expected the Alliance to own the facility. They therefore conceived it as a limited liability company (Davis 2012): a structure that protected investors, since they were not to be the owners of the nuclear waste stored on the site. A private nuclear contractor was to build the facility. The Alliance members expected this model to guarantee them the continued support of the State of New Mexico. To formalize this support, they wanted to convince the state to sign a "host agreement" with the Alliance, supposedly demonstrating to both the Department of Energy and potential private contractors that the state had "given the green light" to the project. Significantly, the ideal business model involved obtaining financing for the project through the Nuclear Waste Fund. Drawing upon the Fund entailed legislative modifications integrating the project within the US nuclear waste policy. If the Alliance succeeded in doing so, then they could depend upon the private nuclear utility companies as their short-term financers and waste providers, while simultaneously making the federal government into the project's long-term guarantor.[24]

In the months that followed the meeting, the Alliance released a call for bids, to which ten private nuclear utilities responded (Davis 2012). In October 2012,[25] AREVA, a French nuclear company at the time,[26] announced that it had won the bid. Since GNEP was launched under the Bush administration, AREVA had been intensively promoting its reprocessing technologies in the United States. A few months before the bid, a high-level AREVA officer had been present at the "National Nuclear Fuel Cycle Summit." During his presentation, he used a stylized image that reassembled several elements of the nuclear fuel cycle in the region of Carlsbad, which was the same image that had been used during the GNEP period to depict AREVA's ambitions in the United States, without designating a specific location. From AREVA's perspective, a consolidated storage facility in Carlsbad would enable them to easily access a very large spent nuclear fuel inventory, which would be transported and stored at the expense of the federal government or other nuclear utilities.[27]

With consolidated storage facility projects, it was no longer necessary to renegotiate the moral, political, technoscientific, and legal terms of the previous resolution of the problem of rendez-vous once reached in Carlsbad around the long-term containment of nuclear waste deep inside the Salado Formation. Nuclear waste, construed as a "market externality," could materially be distanced from its place of production and be stored elsewhere. Consolidated storage facility projects accounted for the concerns of a series of actors, especially those immediately impacted by this "negative market externality": private nuclear utilities whose reactors were reaching the end of their lifetimes with only spent nuclear fuel left to be safely and securely stored on reactor sites; political actors at those sites, impatient to see their lands remediated and available for other uses; and the Department of Energy, concerned about its increased financial liability to private nuclear utility companies. Some of the actors expected these facilities to generate immediate economic and financial profits and/or help them to avoid financial liabilities; others expected them to mediate the evacuation of nuclear waste from their communities and geographies to elsewhere.

Confining spent nuclear fuel inside containers that can be moved around the country, with a price to be paid (as "compensation") to those who provided the storage services and locations, imbued consolidated storage initiatives with the potential to turn nuclear waste into a source of rent. In the words of Kean Birch and Fabian Muniesa, rentiership is based on the appropriation of value through property rights, monopoly conditions, and regulatory or commercial devices and practices (Birch and Muniesa 2020). As a result of these actions, what might be called a new set of economic arrangements was developed.

According to the Nuclear Regulatory Commission (2014), while dry cask storage systems would enhance the immediate safety and security of spent nuclear fuel storage, to remain safe and secure, they also need to be replaced at a hundred-year intervals. Such timescales are not included in debates revolving around consolidated storage facilities, whose licensing periods are currently limited to forty years. Qualified as "temporary," these facilities rely on the assumption that there will be an operating geological repository project in less than a hundred years. The expected end of "interim" storage is not marked on calendar time or on a legal and political agreement; it is a horizon that can continuously be stretched toward far futures, even as benefits and profits accrue to specific private actors in the present whose accountability and liability are limited. In the absence of a repository, it is the present and future local communities who must live with these containers, with no option but to trust the federal government and its agencies to ensure the safety and security of not only those who are currently living there but also those who are to live there in the future[28] (Saraç-Lesavre 2021b).

In this alternative resolution of the problem of rendez-vous, it is neither geological formations, technoscientific institutions, the expertise and knowledge accumulated by them, nor the political arrangements once enshrined into law that are valued. There is no need for long-term negotiations to reach a new political, temporal, and moral arrangement, nor is it necessary to wait for the results of scientific research that examine the "suitability" of a geological formation to contain a given type of waste. Whereas the resolution of the problem of rendez-vous once rested upon the ideal of keeping the present and future generations "free of the trouble" by removing it from the Earth's surface through the mediation of a geological formation that would serve as an interface between human time and geological time, the containerization of spent nuclear fuel rests on different logics. By transforming nuclear waste into transferable and circulable market objects, this alternative mode of containment makes previous resolutions to the problem of rendez-vous—once enshrined in national legislation based on long-term planning, science, bureaucracy, and political decision-making—compete with an approach that relies on markets. Advancing a conceptualization of the market as "a type of distributed collective intelligence mechanism that outperforms planning or democratic decision-making in the task of searching for innovative solutions to uncertain problems" (Frankel et al. 2019), key actors incited major shifts in the discourses and practices of nuclear waste management. In this emerging "container economy," the circulation of nuclear waste (across states and/or globally) would be "sustained by a mobile and disposable workforce, dependent on volatile logistics of infrastructure and nurtured by speculative and asymmetrical geographies" (Leivestad and Markkula 2021). Moreover, surrounding communities and geographies would carry the risk of living with the hazardous remains of the nuclear age, potentially for indeterminate periods of time.

Which Project, Which Values, Which Futures?

These two very different ways of identifying and resolving Carlsbad's problem of rendez-vous held the potential to prepare very different futures not only for the city but also for the country. Yet, the members of the Alliance expected the two futures on the table to be complementary to each other and presented those in continuity. Stored at a consolidated storage facility, spent nuclear fuel would have time to either cool down or be reprocessed and recycled at new nuclear facilities in Carlsbad and its region and, in both cases, become available for disposal at the Salado Formation in an indeterminate future. In seeing and representing things this way, they were guided by specific economic concerns; they construed past investments as assets that were likely to become worthless if they were not used within the lifetime of WIPP.

The nuclear ecology that contained WIPP, which held it together and enabled it to contain waste, depended on the continuity of nuclear activities (Sofia 2000). Those who advocated for consolidated interim storage believed Carlsbad could lose against other "concurrent" YIMBY communities who were or wanted to be part of the nuclear economy. They feared that this fate would befall them based on many potential factors: if their repository, which they proudly referred to as "the only operating deep geological repository," concluded its disposal activities; if the local offices of the National Laboratories and the DOE left Carlsbad; if the specifically designed transportation system degraded; if groups trained for emergency planning around WIPP's transportation routes disappeared; if the city lost its skilled and trained workforce; and, finally, if the national spotlights on Carlsbad were turned off. Then, not only the imagined future but also the investments in Carlsbad's past would have been lost.

WIPP's existing mission, inscribed into law and marked on the regulatory calendar with apparent finitude, made consolidated storage facilities attractive for some as an immediate short-term "solution" that could keep Carlsbad and its surroundings "nuclear" and therefore empowered. These actors believed that this mode of containment could halt the financial liabilities that the breach of contracts incurred to taxpayers, reduce the cost of storage for nuclear utilities, and allow the cleaning up of reactor sites that only have spent nuclear fuel stored on site. This alternative resolution of the problem of rendez-vous was a market arrangement to be reached among the parties of an economic transaction, with its long-term environmental, political, and socioeconomic consequences falling under the responsibility of the federal government and, in the absence of its action, also under the responsibility of those having to live with it, in the present and potentially in the far future.

Not all Task Force members supported this alternative resolution of the problem of rendez-vous, even if they did not publicly voice any dissent. Those who had reservations simply avoided discussions around consolidated storage facility projects; furthermore, not all valued commercial and military nuclear waste equally. One senior local technoscientist referred me to the whole range of US defense nuclear waste that was waiting to be disposed of and passionately explained how the defense waste was the result of collective efforts to ensure national security during the Cold War. He marked the physical containment of this particular type of waste in Carlsbad's deep underground with higher moral and political values than commercial waste. From his perspective, engaging in commercial spent nuclear fuel futures would prevent the city from continuing to contribute to the national interest. It would also, he argued, prevent the city from benefiting from certain advantages associated with integration in the US military complex, which had long protected WIPP from large-scale public scrutiny by placing it under the jurisdiction of national security. He explained,

> WIPP evolved into a defense repository, defense "anything" in this country doesn't get national attention or publicity. . . . We don't plaster billboards of our defense programs. The other thing is, because you realize how much resistance this program [the Yucca Mountain project] attracted, the DOE realized that lots of people will argue about it, will fight about it. Because this mission was for defense waste, to advertise this, to celebrate this, to publicize this process was inviting anybody and everybody who wanted to have a stance against it. So they [the DOE] decided to keep it quiet, do it quietly, do it efficiently, don't make noise, it was our desire to keep it under the radar of those processes, those people who were against, focused on Yucca Mountain. . . . But it was also because of the "defense-only fabric" that buried WIPP in the fabric of the US government.[29]

In a way, he replicates the analysis of German sociologist Georg Simmel on the relationship between secrecy and containment, who notes that "numerous contents of life cannot even emerge in the presence of full publicity. The secret offers, so to speak, the possibility of a second world alongside the manifest world; and the latter is decisively influenced by the former" (Simmel 1950, 330). For this technoscientist, it was important for the Task Force to be discerning regarding the waste they host so that they could continue holding together and therefore containing the conditions that once enabled

WIPP to receive waste. He did not oppose the Alliance's activities, but he did not actively support them either. Not only were there different ways of staying "nuclear," some more valuable and desirable than others, but there were also disagreements about the conditions that enabled Carlsbad to become "nuclear" in the first place. By conceiving of these two modes of containment as complementary to each other, the leaders of the Task Force tried to find ways to hold these potentially conflicting and competing resolutions to the problem of rendez-vous together, even though partially and temporarily.

Conclusion

The city of Carlsbad, its community, geography, and geology, was pregnant with several "slow violences" (Nixon 2011) that had conflicting and competing temporalities. On the one hand, nuclear waste and salt were to become "one" and, in the process, transform the depths of the Salado Formation for hundreds and thousands of years; on the other hand, with the looming possibility that WIPP's disposal activities might conclude, the Task Force members considered that their city was in danger of transforming into a postindustrial ruin. Therefore, the resolution of what the Task Force members identified as their problem of rendez-vous—that of avoiding becoming a postindustrial ruin in the decades to come with the potential end of WIPP's operational phase—relied on their capacity to align their futures with the way different actors identified the problem of rendez-vous. For some, the resolution of the problem of rendez-vous entailed the "permanent" isolation of nuclear waste in the depths of a geological formation and its removal from the biosphere for deep time. For others, it involved the temporary storage of nuclear waste in surface facilities: the integration of nuclear waste in container economies via its containerization.

This chapter has summarized how a group of local actors in Carlsbad anticipated and responded to these competing futures. By tracing their efforts to keep their geography, geology, and community attached to nuclear futures, I have elucidated how local actors *proactively valued* "themselves" and identified "desirable" futures. The valuable and desirable community was made up of individuals who were both "economic agents" and "patriotic citizens," wanting to maximize not only their own utility but also the well-being of the "nation." The intertwined moral, economic, and political logics undergirding their actions were deeply rooted in an entrepreneurial ethos that both built on Carlsbad's long settler colonial history and reflected mainstream political discourses circulating in the contemporary moment. As an entrepreneur community, whose representatives believed in their autonomy and independence, their future was not "something done to them" but instead something they were trying to accomplish "on their own" (Fridman 2014, 96). This constituted a peculiar meeting between a neoliberal pastorate and neoconservatism, which was both encouraged by and encouraging of the transformation of the position and responsibilities of the federal government, which assumed the role of a facilitator, enabler, and guarantor. The traditional themes of responsibility falling under its jurisdiction were being displaced; national nuclear waste futures and the "national interest" were to be carried by voluntary and hybridized individuals and communities, as well as arranged according to the concerns of the market.

But then, what does this say about the evolution of political and technoscientific institutions' initial resolution of the problem of rendez-vous? In an account of the policies revolving around "containment of the atom," Sheila Jasanoff and Sang-Hyun Kim emphasize how local practices have been shaped by sociotechnical imaginaries produced at the "national level" and stressed that "powerful instruments of meaning making and goal selecting often lie within the control of nation states" (Jasanoff and Kim 2009, 124). They insist that it is during the process of goal selections and meaning-making that nations and national interests are produced, with "national imaginations" penetrating in situated practices. They argue that "despite global flows of capital, media, knowledge and skills, the framing and bounding of S&T projects and related policies remain closely intertwined with nation-building" (Jasanoff and Kim 2009, 2015).

The example of Carlsbad shows how the role of the state in constituting the nation and its interests has undergone massive transformations. What local actors said and did indeed highlights how the framing and bounding of science and technology projects remain closely intertwined with the nation and nation-building. The experiences and perspectives of those dealing with nuclear remains, however, also testify to how these projects are not construed, despite "global flows of capital, media, knowledge and skills," but rather reproduce them.

Finally, the containerization of nuclear waste signals the potential abandonment of a resolution of the problem of rendez-vous that once relied on a mode of containment that was to mediate human time and geological time.

By reshaping the political economy of nuclear waste, containerization transforms the temporal orientation of action from the future to the present. It opens up a path to a resolution that relies on the formation of "container economies" that displace risk in time and place, construes nuclear waste as economic objects, and is guided by the morality of the neoliberal economy and its presentist temporal orientation. Therefore, spent nuclear fuel's "permanent" containment deep inside a geological formation and its "temporary" containerization held not only the short-termist and long-termist resolutions of the problem of rendez-vous in tension but also the identification of nuclear waste as a collective or a market concern.

Conclusion

Contemporary industrial societies take various initiatives to act "responsibly" toward the future of humans and the planet by putting in place various arrangements to establish "good" and "desirable" relations across time. They create "Doomsday Vaults" to preserve biodiversity, develop carbon capture and storage projects to displace and re-place the undesirable remains of industrial activities, or reformulate the pricing of water to reduce its consumption and combat drought. They also craft institutions to ensure that "future generations" are politically represented in decision-making processes, imagine living spaces that propel intergenerational solidarity, reconfigure pension schemes and taxation systems to redistribute wealth across time, or pursue urban projects to unbuild cities to make them more relivable both for humans and more-than-humans.

In this book, I have focused on one such highly controversial initiative: the efforts to physically isolate high-level nuclear waste from humans and the environment in the United States. The relevant policy was originally formulated to protect "the present and future generations" from radioactive matter through its confinement in the depths of the Earth. The nuclear waste program was the result of a whole range of political, economic, financial, material, technoscientific, and legal arrangements put in place to collectively produce the physical containment of radioactive matter from humans and the planet. And it is precisely for this reason that the case of nuclear remains helps us understand why it is important to pay close empirical attention to temporalities, responsibilities, and values such arrangements are to hold together and the trajectories they undergo over time, since it can tell us a lot about the societies we live in and values that hold them together (or fail to do so).

As we have seen throughout this book, this kind of analysis requires being attentive to how concerned actors, first, identify temporal gaps in their capacities to act and, second, attempt to overcome them. One way of analyzing such gaps involves paying heed to the formulation of time concepts, which are first and foremost relational (Latour 1997). In the nineteenth century, the advent of capitalism and the ambition to reach a morally and politically desirable time concept around which the capitalist economy could run smoothly resulted in the pursuit of universal time (e.g., Zerubavel 1982; Latour 1997; Galison 2000, 2004). Since the late twentieth century, the increased recognition of relations between human activities and the current state of the planet has provoked the development of the contested concept of the Anthropocene to highlight the entangled relationship between human time and geological time (e.g., Latour 2017; Chakrabarty 2018).

Another approach involves normatively thinking about whether and how relations across *temporally and spatially distanced collectives* can be vested with moral values, especially when these relate to actions that hold the potential to impact people who would live in the distant future (e.g., Meyer 2004; Gosseries 2008; Scheffler 2018). Some ethicists argue that such relations should be treated differently in two key scenarios: when actions impact "future generations" who overlap with current ones (thus transforming a part of the present and future generations into contemporaries), or when actions have negative ramifications not only later but also now (thus transforming the ethical issue in part into a matter of self-interest) (e.g., Butt 2006; Thompson 2009; Gosseries 2023). Others discuss whether there is an ethical duty to leave a "fairer" and more "equal" planet, and whether there are not only obligations for those living in the present but also rights for those living in the future (e.g., Herstein 2008).

A third approach involves engaging with the development of alternative forms of institutions to deal with what some philosophers and political scientists qualify as the short-termism of political institutions: in other words, "democratic myopia" (e.g., MacKenzie 2013; Thompson 2015; Boston 2021a). To that end, some look back and pursue historical analyses of some of the oldest and most durable forms of organizing in an attempt to formulate principles for the conception of "deep time organizations" and ways of coping with challenges related to Earth system governance and planetary stewardship (e.g., Hanusch and Biermann 2020; Hanusch 2023). Just like the one narrated by Kim Stanley Robinson in *Ministry for the Future* (2020)—a fictional UN agency empowered to advocate for and protect the interests of future generations in present-day policy decisions—others explore alternative institutional forms to politically represent the rights of those living in the future (e.g., Smith 2020; Rose 2024) or even new *recognized political entities*—such as land and water entities—whose reidentification as rights-bearing is considered to hold the potential to protect life-worlds of future beings (e.g., Youatt 2017; Tănăsescu 2020).

Adopting a fourth and rather pragmatic approach, in *Nuclear Remains* I addressed the ambition to physically contain radioactive matter for deep time as first and foremost, the result of a network of arrangements. These arrangements were formulated to hold together situated definitions given to desirable and valuable relations between temporally and spatially distanced collectives, and they were designed to do so through negotiations with, among others, materials, institutions, times, technologies, expertise(s), devices, calculations, instruments, communities, geologies, and geographies. I construed all these arrangements as resolutions of problems of rendez-vous, which my work theorizes through and beyond existing scholarly discussions around deep time thinking, generations, and containment.

The notion of the problem of rendez-vous allows us to move away from attempts to attribute morality to a given temporal orientation of action and helps to underline that one can only partially and temporarily succeed in linking temporally and spatially defined collectives, whether action is oriented toward the future or the present. This is because, as an empirical analysis of the quest to geologically contain nuclear remains has shown, values, expectations, and definitions of "responsible action" do not remain immutable. Moving away from normative approaches, following resolutions of specific problems of rendez-vous provides a better understanding of how contemporary societies value "good" relationships between the past, present, and distant futures. And by value, I mean not in the abstract sense, but in the actions that concerned actors take.

While taking "generations" as a unit of analysis may enable establishing specific relations of cooperation and division across times, in *Nuclear Remains* I chose to analytically distance myself from this notion. Instead, I focused on tracing concerned actors' efforts to formulate moral, political, and temporal relations between temporally and spatially defined collectives. Those collectives were neither given nor out there. They resulted from concerned actors' situated efforts to identify and resolve problems of rendez-vous. This analytical choice proved crucial as my empirical cases revealed how "intergenerational" relations are actively articulated through specific arrangements rather than being pregiven categories. The nuclear waste fee system, for instance, didn't simply distribute costs between existing generations but actively constituted specific collectives as "responsible" through the formulation of financial mechanisms.

The nuclear waste policy was arguably one of the very first policy attempts in the United States to act upon an environmental problem with a sense of moral duty to both the present and distant futures. As Bruno Latour rightfully highlights, "To claim that human agency has become the main geological force shaping the face of the Earth, is to immediately raise the question of responsibility" (Latour 2017, 38). However, the characterization of "an epoch through the geolizing of the social"—meaning the acknowledgment of human intervention on Earth—can construct "a monolithic, post-racial 'we.'" In other words, it can fail to account for important historical events or persisting differences in power that impact who is most likely to be affected by what remains of specific human activities. And it can also construct a singular temporality that erases different forms of injustices and produces undifferentiated "histories of responsibilities and reward in geological life" (Yusoff 2018). The case of high-level nuclear waste has been particularly helpful to think about responsibility beyond the establishment of causal relations between historical events (and/or activities) and the current state of geological strata, since its origin, producers, or location are known and delimited.

Tracing resolutions of problems of rendez-vous involves being attentive to many seemingly "boring" and technical details, which, on the one hand, contain contemporary societies' moral and political relations with futures and, on the other, hold the capacity to shut down those relations to public scrutiny since they take place largely out of sight.

Examining resolutions of problems of rendez-vous requires not only inquiring into the overtly visible elements that lend themselves easily to analysis but also investing extra effort in those that need de-scripting. In this book, we have seen the resolution of a problem of rendez-vous take forms as varied as that of a monetary figure, a time concept, a theory of intergenerational justice, a tax, a concrete canister, a geological formation or an engineered infrastructure, and, other times, that of a congressional procedure, an organization, an economic mechanism, a political agreement, a piece of legislation, or a regulatory standard. All these were to hold together and collectively materialize the temporal orientation that policymakers had given to action.

In 2010, what was put on hold was not only the Yucca Mountain project but also the modernist regime of temporality and its orientation toward technoscientific progress and economic growth, along with the ambition of the federal government to take future-oriented action. This suspension reflected broader changes in how US society relates to futures. In *Regimes of Historicity: Presentism and Experiences of Time*, François Hartog (2003) examined moments of change in societies' ways of relating to the past, present, and future, which he named "regimes of historicity." He qualified the period starting from the French Revolution to the fall of the Berlin Wall as the "modern regime of historicity," which he characterized as futurist, and characterized the period starting with the fall of the Berlin Wall as "presentist." According to Hartog, since then, contemporary societies have been caught up in a stagnant present, with no acknowledged past and no real future, since the future constitutes a threat and no longer a promise, making the present gain both an emancipatory and an enclosing character.

Given this temporal shift, what terms of "responsible action" are rendered possible through the meetings between "a stagnant present" and the "long-term epoch" (Haraway 2016)? *Nuclear Remains* has shown that plans to establish "good" and "desirable" relations with far futures have been entangled with "financial times" (Esposito 2018), the temporal dimension of neoliberal reasoning that tends to transform all engagements into economic ones. As the production of long-term oriented action has undergone a whole range of tests, nuclear remains have become increasingly accommodated within short-term oriented plans. The obligations of debt, contract, and compensation, once put in place to establish specific temporal, political, and moral relations, have been reconfigured as the economic logics embedded in them were increasingly prioritized and, therefore, revalued. Efforts to deal with nuclear remains have increasingly joined the promissory expectations of neoliberalism that construe "economic instruments and strategies as the only mechanism through which the future is viewed," making uncertainty "into the constitutional principle of a free society" (Davies 2017, 18), increasingly encouraging members of that "free society," specific temporally and spatially defined collectives, to take the initiative to deal with the looming risks and uncertainties generated by the remains of human activities.

Studying the conception and the career of a nuclear waste program was about studying the values that hold a society together—or fail to do so. For over thirty years, in the United States, the long-negotiated and crafted resolutions to problems of rendez-vous linked the past, present, and future in specific ways. It was neither simply the neoliberal moral economy nor the political system of governance that shifted the temporal orientation of the national nuclear waste policy from the long term to the short term. That shift was both mediated by arrangements formulated to translate a given temporal orientation into action, as well as the trajectories these followed over time under the ever-evolving conditions and expectations of the society.

In analyzing both the efforts and (in)capacities of a society to define what it entails to establish ethical relations between the present and futures, as well as to translate those definitions into action, in *Nuclear Remains*, I pointed out that any attempt to build "good" and "desirable" relationships across time requires concerned actors to continually rearticulate the temporalities, responsibilities, and values they are trying to hold together to ensure that contemporary institutions remain not only au rendez-vous of their existing commitments but also au rendez-vous of what is currently expected of them. This is an extremely difficult task, but also one that is crucial in articulating who "we" are and how "we" care about being au rendez-vous of what remains of our passage on the planet beyond our lifetimes.

Notes

Introduction

1. As anthropology and sociology of waste and discard studies scholars have long highlighted, "waste" is a relational and situated qualification (Hetherington 2004; Reno 2009; Alexander and Reno 2012; Saraç-Lesavre 2015, 2021c; Reno 2016; Besnier and Narotzky 2018; Millar 2018; O'Hare 2019; Reno 2020; Alexander and O'Hare 2023). Whether some nuclear materials are characterized as "waste," "value," "resource," "fuel," or "commodity" are outcomes of situated evaluation and valuation processes that are often contested and temporary (Fourcade 2011; Muniesa 2011; Vatin 2013) and are not their inherent properties. I use nuclear remains to maintain analytical distance from these qualifications and focus on how concerned actors work to define what constitutes "desirable" relations around the handling of these materials, regardless of how they are qualified. Regarding the contested evaluation and valuation processes through which spent nuclear waste gets understood as "waste" or "value," see Saraç-Lesavre (2021c).

2. I use "arrangement" to refer to deliberate assemblages that bring together heterogeneous elements—actors, technologies, institutions, discourses, materials—into provisional configurations (Latour 2005; Akrich, Callon, and Latour 2006). In my case, these arrangements involve inseparable material and semiotic dimensions and require ongoing work to hold together responsibilities, values, and temporalities across time.

3. The amount of spent nuclear fuel stored at nuclear power plants across the country continues to grow by about 2,000 metric tons a year.

4. According to NRC regulations, high-level radioactive waste (HLW) includes (1) spent nuclear fuel, (2) liquid wastes from the main extraction process in fuel reprocessing, and (3) solids made from these liquid wastes (10 CFR 60.2). For the federal nuclear waste program, this meant three main categories of waste to manage: spent nuclear fuel, defense high-level waste, and commercial high-level waste. All of these were to be disposed of at Yucca Mountain. I will differentiate among these categories when necessary for the argument.

5. Social studies of accounting have revealed how accounting practices do far more than simply record financial transactions; they actively shape the social, moral, economic, religious, and political orders they purport to merely measure. Accounting conventions establish what counts as valuable, who bears responsibility, and how obligations are distributed across time and space (see Miller 1998; Quattrone 2004; Ezzamel 2009; Spence 2010; Deringer 2018, 2024).

6. Hale's definition of "long-problems," as characterized by temporal distance between causes and effects, mischaracterizes how temporal challenges work (Hale 2024). Consider nuclear waste: while it poses both immediate radiological hazards and long-term contamination risks, the effects of exposure are immediate and potentially deadly; there is no temporal distance between cause and effect. What makes nuclear waste a temporal challenge is not distant effects, but the effort to formulate arrangements to produce the temporal orientation given to responsible action across time. The temporal complexity lies in governance arrangements, not in causal distance.

7. For an excellent discussion on "technocratic myopia," see White (2024).

8. Social scientists have long acknowledged that "time" is not a neutral backdrop against which social life unfolds but is actively produced through situated sociomaterial practices and relationships and that it produces practices and relationships in return. From Evans-Pritchard's (1939) analysis of how Nuer cattle-herding rhythms shape temporal categories, to Bourdieu's (1990) exploration of how Kabyle agricultural practices generate distinct temporal rhythms, to Munn's (1992) ethnography of how Gawan exchange practices create intersubjective time, scholars have demonstrated that time itself emerges from sociomaterial relations rather than preceding them (see also Malinowski 1990; Hassard 1990; Quattrone 2005; Galison 2014; Gell 2021). This relational understanding of time is crucial for analyzing how societies operate: through the production and maintenance of temporal relations.

9. Since the 1990s, nuclear waste warning markers have fascinated historians, anthropologists, semiologists, and archaeologists as a unique challenge in communicating across deep time. These permanent markers, designed to warn future civilizations about buried radioactive dangers for millennia, raise fundamental questions about not only communication, cultural continuity, and the limits of semiotic systems but also about who we are as "humans." Some scholars have actively contributed to marker design, while others have analyzed these efforts as revealing broader tensions about how societies attempt to project their definitions of responsible action into uncertain futures (Sebeok 1984; Kaplan and Adams 1986; Hora and Von Winterfeldt 1997; Van Wyck 1997; Benford 2000; Moisey 2012; Holtorf and Högberg 2015; Galison and Moss 2015; Moisey 2017; Joyce 2020, 2024).

10. Social scientists pay attention to how containment, as enclosure, mediates the production of domestic, political, moral, planetary and economic orders (e.g., Chrulew 2017; Keck 2017; Radin and Kowal 2017; Newell 2018; Parker 2023, 2025). Examining

the role of containment technologies in reconfiguring space/time relations, several scholars have highlighted the part they play in the growth and transformation of logistics and emphasized their importance in the maintenance and expansion of capitalist order (e.g., Chua et al. 2018; Danyluk 2018; Coe 2020; Banoub and Martin 2020). In a related vein, others have highlighted that this involves compressing temporal processes into spatial arrangements while extending spatial relations across time in ways that enable its production and maintenance (e.g., Simpson 2019; Randle 2024). While these works are crucial, they often miss how containment technologies work to hold together heterogeneous elements across time and space and fail to recognize how this holding together constitutes an active, experimental method for political, economic, and moral order-making. In my analysis, I treat storage, disposal, and concealment as interconnected forms of containment that resolve problems of rendez-vous through the inscription of temporalities, responsibilities, and values.

Chapter 1

1. National Association of Regulatory Utility Commissioners v. DOE, 2010 U.S. App. LEXIS 25579 (D.C. Cir. December 13, 2010).

2. For an outstanding analysis of how a "reference person" was identified and what that identification included and excluded, see Cram (2015).

3. Of course, one cannot help but note the irony of the absence of West Africa in the nuclear order, while its people and underground have long been used as resources the industry could extract from. For an in-depth analysis of the production and maintenance of nuclear order, see Hecht (2012).

4. There have been many rich and critical accounts of the history of nuclear waste management in the United States (see, e.g., Shrader-Frechette 1993; Flynn 1995; Macfarlane and Ewing 2006; Hamblin 2008; Walker 2009; Schwartz 2011; Stewart and Stewart 2011; Alley and Alley 2013).

5. The DOE was formed in October 1977, following the dissolution of the Energy Research and Development Administration (ERDA) by President Carter. The Office of Waste Isolation (OWI) had been managed for ERDA by the Union Carbide Corporation's Nuclear Division in Oak Ridge, Tennessee, which was in charge of commercial spent nuclear fuel. The OWI was created within the context of a program designed for the management of defense and commercial nuclear waste by the ERDA in February 1976 (see Claiborne 1977).

6. In 2005, the regulatory period of containment was extended to a million years (Shrader-Frechette 2005).

7. Far from a mere technical distinction, the choice between taxes and fees reflects deep disagreements about whether governments can be trusted with broad fiscal powers or must be constrained through user-pay mechanisms (Buchanan and Musgrave 1999).

This theoretical debate gained concrete urgency with environmental policy's "polluter pays principle" with legal scholars like Sanford E. Gaines (1991) analyzing how different fiscal instruments generate different political and distributive outcomes. Sociologist Rémi Barbier (2005) has shown how different fiscal systems produce distinct distributive effects and conflicts in environmental policy contexts. More recently, anthropologists have examined how fiscal mechanisms embed specific ethical logics about reciprocity, obligation, and state–citizen relations and demonstrated how those respond to fundamental moral questions about responsibility and social membership, not merely technical considerations (Ballestero 2015; Venkatesan 2020; Johannson 2020; see also Makovicky and Smith 2023; Mugler et al. 2024).

8. This approach reflected the broader ascendance of the "polluter pays principle" (Gaines 1991) in environmental policy, which was codified in US law through the Comprehensive Environmental Response, Compensation, and Liability Act in 1980 (CERCLA). For a long history of the polluter pays principle, see Fressoz (2012).

9. H.R. 5016 1981.

10. Standard Contract for Disposal of Spent Nuclear Fuel and/or High-Level Radioactive Waste, 48 Fed. Reg. 16. 599 (April 18, 1983).

11. Burn-up measures how much nuclear fuel has been consumed in a reactor. It is essentially how "used up" the fuel is. It can be expressed as the percentage of fuel atoms that have undergone fission or the amount of energy produced per unit weight of fuel in the reactor.

12. The DOE's interpretation of the term "electricity generated and sold" in the contract resulted in lawsuits against the DOE. The final court order redefined the spent nuclear fuel for which the industry was financially responsible as only the electricity sold through commercial transactions between the utilities and their ratepayers. See Consolidated Edison v. U.S. D.O.E. 870 F. 2d 694.

13. By April 30, 1983, the DOE had signed seventy bilateral standard contracts with fifty-six parties, including forty-six nuclear utilities (accounting for sixty contracts due to multiple plant operations), eight owners of industrial test reactors, and two nuclear fuel vendors (see U.S. Government Accounting Office, Comptroller General 1986).

14. United States Court of Appeals, No. 11–1066, National Association of Regulatory Commissioners v. United States Department of Energy, Argued September 25, 2013 and Decided November 19, 2013.

Chapter 2

1. According to Lisa Heinzerling, in 1965, the economist Kenneth Boulding testified before the Congressional Subcommittee on Irrigation & Reclamation and recited the poem (Heinzerling 1999).

2. Queen Elizabeth COP26 Speech. October 26, 2021. https://www.youtube.com/watch?app=desktop&v=eXvfqUe4EFQ (last visited December 23, 2021).

3. Several crucial studies have described the historical origins, evolution, and the wide range of uses of discounting (Miller 1998; Muniesa et al. 2017; Deringer 2017, 2018, 2024). Some have shown the vital role they play in the making of futures (Deringer 2017; Doganova 2018a, 2024), and others have demonstrated the political and moral values enacted in them (Vatin 2008; Spence 2010; Saraç-Lesavre 2015, 2018; Doganova 2018a; Deringer 2018, 2024).

4. A rich body of scholarship has shown that economics and morality are deeply entangled rather than separate spheres. Economic theories, conventions, and practices don't simply reflect preexisting moral orders but actively shape what counts as valuable, who deserves what, and how social relationships should be organized (e.g., Thompson 1971; Vatin 2007; Browne 2008; Maurer 2008; Trompette and Griffiths 2011; Fassin and Eideliman 2014; Zelizer 2017).

5. The work of Kip Viscusi, a neoclassical economist at Vanderbilt Law School, provides a good illustration of this perspective (see, e.g., Viscusi and Zeckhauser 1976; Moore and Viscusi 1990; Viscusi 2000, 2006; Viscusi and Aldy 2003).

6. For arguments against the use of long-term discounting in environmental and policy contexts, particularly regarding its ethical implications for future generations, see Goodin (1978), Ackerman and Heinzerling (2002), Heinzerling (2001, 2006), Rowell (2010), and Sunstein and Rowell (2007).

7. This expansion generated scholarly debates. Legal scholar Edward Morrison (1998) describes the evolution as follows: "Although previous administrations issued executive orders encouraging agencies to consider the economic impact of proposed regulations, President Reagan's executive order, EO 12291, 46 Fed Reg 13193, was the first to require cost-benefit analysis. Section 2 of EO 12291 required agencies to ensure that the social benefits of a proposed regulation exceed its social costs. In 1993, President Clinton issued the Executive Order 12866, 58 Fed Reg 51735 (1993), which generally affirms the approach of the Reagan order. Unlike Reagan's order, however, Executive Order 12866 § 1(b) merely endorses cost-benefit analysis as a tool for evaluating regulatory options and does not require that benefits outweigh costs. 58 Fed Reg at 51735–36" (1333).

8. These figures are referred to as "internal rates" and are presented inside a specific section of the OMB's Circular A-94.

9. Field notes from discussions with retired technoscientists working on the Yucca Mountain project at the Office of Civilian Radioactive Waste Management, Summerlin, Nevada, April 2012.

10. Field notes from discussions with retired technoscientists working on the Yucca Mountain project at the Office of Civilian Radioactive Waste Management, Summerlin, Nevada, April 2012.

11. Field notes from a discussion with a retired OCRWM DOE technoscientist, February 2012, Phoenix, Arizona.

12. Weitzman (2001) first conducted survey research on a large number of PhD economists and on a more restricted number of prominent economists. Rather than asking the question directed at a holistic "future," Weitzman separated the future into four distinct, temporal categories and asked the respondents to name what discount rate they would suggest for each category. The economists' answers showed a very significant uncertainty about long-term discounting: the mean of the discount rate was around 4 percent, whereas its standard deviation was around 3 percent. Weitzman concluded that discount rates for long periods of time are uncertain. Departing from this analysis, Weitzman created a model in which he attributed the same probability to ten different yet possible discount rates in the future. This model produced a declining discount rate over time, which resulted in a very weak discount rate. He called this phenomenon "gamma discounting" (also called declining and hyperbolic discount rates): a declining discount rate that results in discounting the long-term future at the lowest level. To calculate the discount rate, Weitzman allocated risks to present generations.

13. In its 2006 report on cost-benefit analysis, the OECD began to promote the work of Weitzman as a way of addressing the tension between high and low discount rates. National commissions in the United Kingdom and France on long-term discounting for regulatory analysis endorsed declining discount rates and recommended their adoption by their respective governments. In September 2011, the US Environmental Protection Agency (EPA) formed an expert panel on long-term discounting to address the question: "How should benefits and costs be discounted in an intergenerational context?" The panel recommended that the EPA use declining discount rates for environmental regulations ("Resources for the Future," http://www.rff.org/files/sharepoint/WorkImages/Download/RFF-DP-12-53.pdf).

14. Interview with a retired Yucca Mountain project manager, Las Vegas, Nevada, April 2012.

Chapter 3

1. Interview with Steve Frishman and Judy Treichel, leaders of Nevada's opposition to the Yucca Mountain Project, Washington, DC, May 2011, and Summerlin, Nevada, April 2012.

2. Interview with Steve Frishman and Judy Treichel, Washington, DC, May 2011, and Summerlin, Nevada, April 2012.

3. Interview with Steve Frishman and Judy Treichel, anti-Yucca Mountain activist, Washington, DC, May 2011.

4. Until its dissolution in 1995, the OTA was an office of the US Congress in charge of the evaluation of scientific and technological decisions. It provided congressional members and committees with alternative analyses of scientific and technical issues.

5. The OTA's analysis of the management and financing of the nuclear waste policy was based on a study that the OTA specifically requested from the Congressional Budget Office. The Congressional Budget Office is a staff institution that is independent from the committees and parties, and it mainly provides technical knowledge produced by economists who are employed by the Office.

6. The Committee on Energy and Natural Resources and the Subcommittee on Nuclear Regulation of the Committee on Environment and Public Works. Joint Hearing. First Session on S. 637 and S.1662, 97th Cong. 286. 1981. Testimony of Thomas A. Cotton.

7. Submitted to the Joint Hearing of the Committee on Energy and Natural Resources and the Subcommittee of the Committee on Environment and Public Works. First Session on S. 637 and S.1662. October 5, 1981.

8. Several studies conducted during this period reached the same conclusion (see, e.g., Interagency Review Group 1978; US Congressional Budget Office 1982; Office of Technology Assessment 1982; US Department of Energy 1984).

9. Legislative bills introduced at the House concerning the national nuclear waste policy that include the mention of a trust fund in the 97th Congress are H.R. 2840 and H.R. 1993.

10. John Hezir, a former expert at the OMB, notes that various CBO cost estimates referred to the Fund as a trust fund and provided Congress with cost estimates. These included projected surplus or deficit balances for the Fund, on the assumption that monies would be spent from the Fund as needed to meet program schedules and that the Fund could borrow funds from the Treasury to cover any short-term requirements not covered by fees (Hezir 2011).

11. The Committee on Energy and Natural Resources and the Subcommittee on Nuclear Regulation of the Committee on Environment and Public Works. Joint Hearing. First Session on S. 637 and S.1662, 97th Cong. 286 (1981). Testimony of Thomas A. Cotton.

12. Nuclear Waste Policy Act of 1982, 42 U.S.C. § 10101–10270 (2006).

13. Apportionment is a budgetary measure that can affect the timing of the obligation of appropriated funds (see Schick 2000, 288).

14. By April 30, 1983, the DOE had signed an initial seventy contracts with fifty-six different organizations, including forty-six nuclear utilities (accounting for sixty contracts due to multiple plant operations), eight owners of industrial test reactors, and two nuclear fuel vendors. These contracts covered all of the nation's then commercial generators and owners of spent nuclear fuel. The contracts set forth the specific terms and conditions, as well as the procedures for collection and payment of fees under which the DOE should make available disposal services for commercial spent nuclear fuel under the act.

15. DOE Standard Contract with utilities as finalized in 48 Fed. Reg. 5458.

16. Carrying out the act's contracting provisions, the DOE first drafted a standard contract and published it as a proposed rule in the *Federal Register* on February 4, 1983. On March 3, 1983, the DOE held public hearings on the proposed contract in Washington, DC. In addition to the oral comments at the public hearings, the DOE received written comments from eighty-five organizations representing electric power associations, nuclear power companies, environmental organizations, consumer protection associations, state and federal agencies, and individual citizens.

17. Nuclear Waste Policy Act of 1982, 42 U.S.C. § 10101–10270 (2006).

18. Note from interview with Jane, Nevada, April 2012.

19. Hezir uses this expression to designate the transformation of the Fund in use.

20. Notes from an interview with a retired OCRWM project manager, Nevada, April 2012.

21. In the president's fiscal year 1995 budget, the DOE proposed that the Fund should become a mandatory spending fund to be directly spent without appropriations. In 1997, Republican Representative Upton proposed a bill in the House that sought to transform the fee into an annual user fee (H.R. 1270), maintaining it as subject to appropriations. In 1997, Republican Senator Murkowski (S.104) proposed a two-part fee, suggesting a modification of the standard contract and a "deferred fee" in the Senate. With this modification, the utilities could retain the fund that is not appropriated during the same year and pay the rest once the repository starts accepting waste, plus the interest. In 1998, the DOE proposed changing the standard contract to allow utilities to retain the fund that exceeded the level of that year's appropriation. In 1999, a House bill submitted by Republican Representative Upton (H.R. 45) suggested that the Fund should become off-budget. In 2001, the DOE's Panel on the Alternative Means to Financing and the Management of the Nuclear Waste Program proposed that the Fund should become a mandatory spending fund, but one that is subject to appropriations, making it part of appropriated entitlements but not subject to discretionary spending. In 2004, Republican Representative Barton proposed a bill (H.R. 3981) seeking to reclassify the Fund revenues as offsetting collections, and in 2005, in its 2006 Budget request, the DOE re-proposed the same.

Chapter 4

1. Statement of Reasons Supporting the Governor of Nevada's Notice of Disapproval of the Proposed Yucca Mountain Project. Kenny C. Guinn. April 8, 2002. http://www.yuccamountain.org/govveto0402.htm.

2. Those efforts had already begun in the 1990s. Here, I list some of the proposals that the DOE or congressional representatives submitted to Congress throughout the career of the Yucca Mountain project to render the Fund more accessible. In the president's

budget for the fiscal year 1995, the DOE proposed legislation authorizing 50 percent of annual fees to be spent directly without appropriations and to transform the spending for the nuclear waste program into "mandatory spending." In the president's budget for fiscal year 1996, the DOE proposed legislation to reclassify the nuclear waste fund spending as a mandatory account and to liberate it from annual appropriations. In 1997, a House bill (H.R. 1270), introduced by Republican Representative Fred Upton from Michigan and cosponsored by 160 congressional representatives, proposed amending the Nuclear Waste Policy Act to allow annual appropriations from the Fund to exceed its annual receipts until the opening of the repository, while keeping the spending subject to annual appropriations. In 1997, a Senate bill (S.104), introduced by Republican Senator Frank Murkowski from Alaska and cosponsored by thirty other senators, proposed setting up a two-part fee that aimed at "offsetting collections," limiting fee collection to annual appropriations, and keeping the spending subject to annual appropriations. In 1998, the DOE proposed modifying the standard contracts to allow utilities to retain the portion of the fee that exceeded the annual appropriations and make utilities pay the deferred fees plus interest upon waste acceptance. In 2001, the DOE set up an Experts Panel and explored "Alternatives Means for Financing and Managing" its nuclear waste program. The Panel recommended that the Nuclear Waste Fund should become a part of "Appropriated Entitlements," which entailed transforming the Fund into a mandatory spending fund and keeping it subject to direct appropriations. In 2004, the president's budget for the fiscal years 2005 and 2006 also proposed that the Fund should become subject to offsetting collections. For a review of those proposals, see US Department of Energy (2001), "Alternative Means of Financing and Managing the Civilian Radioactive Waste Management Program," DOE/RW-0546 and Hezir (2011).

For the legislative proposals, see Nuclear Waste Policy Act of 1997, S. 104, 105th Cong. 1997 (unenacted). Nuclear Waste Policy Act of 1997, H.R. 1270, 105th Cong. 1997 (unenacted). Nuclear Waste Policy Act of 1999, H.R. 45, 106th Cong. 1999 (unenacted). Energy and Water Development Appropriations Act, 2005, H.R. 4614, 108th Cong. 2004 (unenacted). A Review of the Department of Energy's Yucca Mountain Project, and Proposed Legislations to Alter the Nuclear Waste Trust Fund: Hearing Before the Subcommittee on Energy and Air Quality of the Committee on Energy and Commerce House of Representatives, 108th Cong. 1. 2004. To reclassify fees paid into the Nuclear Waste Fund as offsetting collections, and for other purposes, H.R. 3981, 108th Cong. 2004 (unenacted).

3. A Review of the Department of Energy's Yucca Mountain Project, and Proposed Legislations to Alter the Nuclear Waste Trust Fund: Hearing before the Subcommittee on Energy and Air Quality of the Committee on Energy and Commerce, House of Representatives, 108th Cong. 1. 2004.

4. A Review of the Department of Energy's Yucca Mountain Project, and Proposed Legislations to Alter the Nuclear Waste Trust Fund: Hearing before the Subcommittee

on Energy and Air Quality of the Committee on Energy and Commerce House of Representatives, 108th Cong. 1. 2004. To reclassify fees paid into the Nuclear Waste Fund as offsetting collections, and for other purposes, H.R. 3981, 108th Cong. (2004) (unenacted).

5. DOE Standard Contract with utilities as finalized in 48 Fed. Reg. 5458.

6. A Review of the Department of Energy's Yucca Mountain Project, and Proposed Legislations to Alter the Nuclear Waste Trust Fund: Hearing before the Subcommittee on Energy and Air Quality of the Committee on Energy and Commerce House of Representatives, 108th Cong. 1. 2004.

7. A Review of the Department of Energy's Yucca Mountain Project, and Proposed Legislations to Alter the Nuclear Waste Trust Fund: Hearing before the Subcommittee on Energy and Air Quality of the Committee on Energy and Commerce House of Representatives, 108th Cong. 1. 2004. To reclassify fees paid into the Nuclear Waste Fund as offsetting collections, and for other purposes, H.R. 3981, 108th Cong. (2004) (unenacted).

8. Re: Yucca Mountain project budget in FY04 Energy and Water Appropriations Bill. November 4, 2003. Letter from Alliance for Nuclear Accountability * American Public Health Association * Clean Water Action * Environmental Working Group * Friends of the Earth * Greenpeace Honor the Earth * League of United Latin American Citizens * National Environmental Trust * Nuclear Information & Resource Service * Peace Action * Physicians for Social Responsibility * Public Citizen's Critical Mass Energy and Environment Program * Sierra Club * Taxpayers for Common Sense Action * 20/20 Vision * US Public Interest Research Group * The Wilderness Society * Women's Action for New Directions * Women Legislators' Lobby (WiLL).

9. Information restored based on the official directory of the US Congress. See congressional directory from 105th Congress through 110th Congress. http://www.gpo.gov/fdsys/browse/collection.action?collectionCode=CDIR&browsePath=107%2F2001-12-07%2FF1&isCollapsed=false&leafLevelBrowse=false&isDocumentResults=true&ycord=416

10. Energy Policy Act of 2005, Pub. L. No. 109–58, 119 Stat. 594 (2005).

11. S.2099 introduced in the Senate Committee on Environment and Public Works on December 14, 2005, 109th Congress, 1st Session.

12. Fiscal Year 2007 Energy and Water Appropriations Act. S. Rep No. 109–274, at 1. 2007.

13. S.3962 introduced in the Senate at the Committee on Energy and Natural Resources on September 27, 2006, 109th Congress, 2nd Session.

14. S.784 introduced in the Senate at the Committee on Environment and Public Works on March 6, 2007, 110th Congress, 1st Session.

15. Energy and Water Development Appropriations Act, 2002, Pub. L. No. 107–66. 2001.

Consolidated Appropriations Resolution, 2003, Pub. L. No. 108–7, 117 Stat. 11. 2003.

Energy and Water Development Appropriations Act, 2004, Pub. L. No. 109–103. 2003.

Consolidated Appropriations Act, 2005, Pub. L. No. 108–447. 2004.

Energy and Water Development Appropriations Act, 2006, Pub. L. No. 109–103. 2005.

Consolidated Appropriations Act, 2008, Pub. L. No. 110–161. 2007.

Making Further Continuing Appropriations for Fiscal Year 2009, and for Other Purposes, Pub. L. No. 111–6. 2009.

Energy and Water Development and Related Agencies Appropriations Act, 2010, Pub. L. No. 111–85, 123 Stat. 2845. 2009.

16. Notes from a discussion with Jane, Nevada, April 2012.

17. Interview with Jane, a retired OCRWM manager and a geologist, Nevada, April 2012.

18. Interview with Jane, a retired OCRWM manager and a geologist, Nevada, April 2012.

19. Interview with Jane, a retired OCRWM manager and a geologist, Nevada, April 2012.

20. Information on the NRC website about the Yucca Mountain Licensing Review. http://www.nrc.gov/waste/hlw-disposal/yucca-lic-app.html.

21. A swing state is a state whose electoral color is open to change, and winning the elections in swing states is important, especially in close elections, which results in the presidential candidates putting a lot of effort into fulfilling the demands of those states. As importantly, since US presidents are not elected by the popular vote but by the Electoral College, candidates spend most of their time campaigning in the states that they think they need to win to secure the required 270 electoral votes. Several newspaper articles from the *New York Times* underline the importance of swing states for both the 2008 and the 2012 presidential elections, one of which is Nevada. See, for example, the following:

Bostock, Mike, and Shan Carter. 2012. "Swing History." *New York Times*, October 15.

Cooper, M. 2012. "Swing States Key to Election Are Mixed Lot." *New York Times*, May 5.

Steinhauer, Jennifer. 2008. "Democrats Rush to Build Up Campaigns in Nevada." *New York Times*, January 10.

22. According to the spokesperson of the opposition to the Yucca Mountain project in Nevada, politicians who supported the project quickly characterized their opposition

as a "Not in My Back Yard" movement. The resistance of Nevada was characterized as "egoistic, self-interested," as a refusal to sacrifice their private interests for the "wellbeing of the nation" and for the "national interest." To that end, their local economy was qualified as immoral through the accentuation of its reliance on gambling and prostitution, while the disposal project was vested with a higher morality and qualified as one that served "the common good." Interview with Steve Frishman and Judy Treichel, Summerlin, Nevada, April 2012.

23. The *New York Times* describes Harry Reid's Senate election campaign as follows: "The Senator has championed the closing of Yucca Mountain as a nuclear waste repository, an enormously popular stance back home but also one that has permitted him to flex what he described in the interview as two other leading points in his election campaign: the cost of the state's losing a majority leader, and his relationship with Mr. Obama, who pledged during the 2008 race to abandon the Yucca repository plans" (Nagourney 2009).

24. Notes from the open meeting of the Commission in March 2010. The Secretary of Energy made an opening statement, during which he made clear that the agenda of the Commission excluded any discussion on the Yucca Mountain project's failure or success.

25. Webpage of the *New York Times* dedicated to the 2010 Senate elections, accessible at http://elections.nytimes.com/2010/results/senate.

26. US Department of Energy. 2010. "Department of Energy FY 2011 Congressional Budget Request." Office of Chief Financial Officer webpages, February 1. http://energy.gov/cfo/downloads/fy-2011-budget-justification.

27. US Department of Energy. 2010. "Department of Energy FY 2011 Congressional Budget Request." Office of Chief Financial Officer webpages, February 1. http://energy.gov/cfo/downloads/fy-2011-budget-justification.

28. US Department of Energy. 2011. "Department of Energy FY 2012 Congressional Budget Request." Office of Chief Financial Officer webpages, February 1. http://energy.gov/cfo/downloads/fy-2012-budget-justification.

US Department of Energy. 2012. "Department of Energy FY 2013 Congressional Budget Request." Office of Chief Financial Officer webpages, January 20. https://www.energy.gov/cfo/downloads/fy-2013-budget-justification.

29. Energy and Water Development and Related Agencies Appropriations Act, 2012, H.R. 2354, 112th Cong. 2011 (unenacted).

30. Interview with a retired OCRWM scientist, Nevada, April 2012.

31. Email communication with a retired OCRWM scientist, May 2013.

32. Interview with a retired DOE scientist who worked for the Yucca Mountain project, Arizona, March 2012.

33. Many other press articles underline the same points with analogies to a biological end of life:

Wald, Matthew L. 2012. "Is Yucca Mountain Still Dead?" *New York Times*, May 3.

Wald, Matthew L. 2011. "How Dead Is Yucca Mountain?" *New York Times*, December 9.

Northey, Hannah. 2011. "GAO: Death of Yucca Mountain Caused by Political Maneuvering." *New York Times*, May 10.

"Yucca Mountain Is Still Alive." 2012. *Pahrump Valley Times*. http://pvtimes.com/news/yucca-mountain-still-alive-consultants-predict/.

34. Interview with Steve Frishman and Judy Treichel, Summerlin, Nevada, April 2012.

35. Notes from an interview with a retired OCRWM manager and technoscientist, Nevada, April 2012.

36. Interview with Jane, Nevada, April 2012.

37. The description of the Board is available from the official website of the NRC: http://www.nrc.gov/about-nrc/organization/aslbpfuncdesc.html (last visited April 2, 2014).

38. "Who Is Responsible for Nuclear Safety?" n.d. Union of Concerned Scientists. http://www.ucsusa.org/nuclear_power/making-nuclear-power-safer/who-is-responsible/the-nuclear-regulatory-commission.html.

39. Interview with Jane, April 2012, Nevada.

40. See the two reports released by two Committees:

Majority Staff Report, Committee on Oversight and Government Reform. 2011. "A Crisis of Leadership: How the Actions of Chairman Gregory Jaczko Are Damaging the Nuclear Regulatory Commission."

Majority Staff Report, Committee on Science, Space and Technology. 2011. "Yucca Mountain: The Administration's Impact on U.S. Nuclear Waste Management Policy."

41. Majority Staff Report, Committee on Science, Space and Technology. 2011. "Yucca Mountain: The Administration's Impact on U.S. Nuclear Waste Management Policy."

42. According to Todd Garvey, a legal attorney at the Congressional Research Service, following the inconclusive results of the first lawsuits filed against the DOE questioning its authority to submit the motion, it was the concurring opinion of the two circuit judges, Brown and Kavanaugh, that had reoriented the lawsuits against the NRC. In their concurring opinions, they noted that the NRC had the final word on whether the executive branch had the authority to terminate the project and pointed out that the petitioners' focus on the president and the DOE was not efficient. For Garvey, this resulted in the reorientation of the lawsuits concerning the motion against the NRC (see Congressional Research Service 2012).

43. In Re: Aiken County, et al. On Petition for Writ of Mandamus, No. 11–1271, Argued on May 2, 2012, and decided in August 2013. http://www.cadc.uscourts.gov/internet/opinions.nsf/BAE0CF34F762EBD985257BC6004DEB18/$file/11-1271-1451347.pdf.

44. The Savannah River Site, one of the three major federal sites where defense high-level waste is stored in the United States, is partially located in Aiken County.

45. Located along the Mississippi River in Minnesota, the Prairie Island Indian Community was one of the communities physically closest to both a nuclear power plant and temporary nuclear waste storage. Community members were concerned about Xcel Energy's aging forty-year-old Prairie Island Nuclear Generating Plant and its twenty-nine dry casks storing spent nuclear fuel, since they were located on Mdewakanton Dakota ancestral land just 600 yards from tribal members' homes. See the description available on the Prairie Island Community website: http://www.prairieisland.org/community/nuclear-power/.

46. NARUC is the national association representing the State Public Service Commissioners who have regulated essential utility services in the United States since 1889.

47. In Re: Aiken County, et al. On Petition for Writ of Mandamus, No. 11–1271, Argued on May 2, 2012, and decided in August 2013. http://www.cadc.uscourts.gov/internet/opinions.nsf/BAE0CF34F762EBD985257BC6004DEB18/$file/11-1271-1451347.pdf.

48. In Re: Aiken County, et al. On Petition for Writ of Mandamus, No. 11–1271, Argued on May 2, 2012, and decided in August 2013. http://www.cadc.uscourts.gov/internet/opinions.nsf/BAE0CF34F762EBD985257BC6004DEB18/$file/11-1271-1451347.pdf.

49. In Re: Aiken County, et al. On Petition for Writ of Mandamus, No. 11–1271, Argued on May 2, 2012, and decided in August 2013. http://www.cadc.uscourts.gov/internet/opinions.nsf/BAE0CF34F762EBD985257BC6004DEB18/$file/11-1271-1451347.pdf.

Chapter 5

1. Legally, transuranic waste is defined in the Waste Isolation Pilot Plant Land Withdrawal Act (LWA) as "waste containing more than 100 nanocuries of alpha-emitting transuranic isotopes per gram of waste, with half-lives greater than 20 years, except for (A) high-level radioactive waste; (B) waste that the Secretary of Energy has determined, with the concurrence of the Administrator [of the Environmental Protection Agency], does not need the degree of isolation required by the disposal regulations; or (C) waste that the Nuclear Regulatory Commission has approved for disposal on a case-by-case basis in accordance with [10 C.F.R. Part 61]." Pub L. No. 102-579, § 2(20), 106 Stat. 4777, 4779 (1992). It is important to note here that although the formal definition is highly technical, it is extremely open to revaluation in that it doesn't specify their origin, source, or location.

2. For cultural, historical, and socioeconomic transformations that the cradle-to-grave nuclear economy brought to New Mexico, see Nash (1990), Masco (2006), Voyles (2015), and Genay (2019).

3. There is increased interest in examining communities whose livelihoods rely on local nuclear economies (e.g., Kalshoven 2022; Bell 2024).

4. Interview with the Carlsbad mayor, Carlsbad, New Mexico, April 2012.

5. Infrastructures require maintenance; their sheer existence doesn't mean they remain "assets" to capitalize on; in June 2025, the Government Accountability Office published a study describing the degraded condition of WIPP's infrastructure (GAO 2025).

6. Notes from the former mayor's presentation and speech at the WM 2012 Conference, Panel Session 105, Closing Featured Session US Administration's Blue Ribbon Commission on America's Nuclear Future, Phoenix, Arizona, March 1, 2012.

7. For scholarship that has highlighted environmental injustices generated by industrial activities and their remains (e.g., Nixon 2011; Gordillo 2014; Stoler 2016; Bubandt and Tsing 2018; Liboiron 2021).

8. My initial fieldwork took place during ethnographic field trips in the winter and spring of 2012/13. I conducted semi-structured interviews with prominent actors involved in the activities of the Nuclear Task Force and undertook hundreds of hours of informal discussions, in addition to direct observations of Task Force meetings, events, and hearings organized and/or attended by Task Force members. The events included the *National Nuclear Fuel Cycle Summit* in Carlsbad and the WM2012 Conference in Phoenix. Additionally, my research draws on transcripts of various public hearings held by the DOE and the Blue Ribbon Commission on America's Nuclear Future in Carlsbad and its region, along with institutional documents collected from the DOE's Carlsbad field office. I kept in touch with key interlocutors, continuing to exchange information and meet periodically through 2021.

9. I use the names of public figures and officials when citing positions that they had already stated publicly in their formal capacities. I have not anonymized my key interlocutor, Abe Van Luik, who passed away in 2016. I retain his name as acknowledgment for his invaluable help.

10. Interview with a senior LANL scientist, Carlsbad, New Mexico, April 2012.

11. Interview with a senior DOE Carlsbad field office scientist, Carlsbad, New Mexico, April 2012.

12. Interview with Carlsbad's mayor, Carlsbad, New Mexico, April 2012.

13. Noted during interviews and discussions with several local residents throughout the fieldwork.

14. Comments by former New Mexico Senator Domenici, member of the Commission on America's Nuclear Future, Carlsbad, New Mexico. Official transcript, January 27, 2011 (see also Blue Ribbon Commission on America's Nuclear Future 2012).

15. In February 2011, the Commission held a public hearing with a group of social scientists who underlined the importance of the "participatory turn" in siting decisions.

16. In 2014 and 2021, the US Department of Energy launched federally funded research programs to explore the possibility of "consent-based siting" for nuclear facilities. These programs investigate how such an approach might be implemented, examining questions of how "consent" gets defined, measured, and operationalized in practice. The research analyzes potential mechanisms for consent-based siting, the temporal dimensions of community agreement, and the practical challenges of translating consent into institutional arrangements. For analysis of these research programs and their findings, see Richter et al. (2022), Bell (2023), Tuler and Webler (2023), Gupta et al. (2024), and Uribe et al. (2024).

17. Notes from discussions with several Task Force members.

18. Notes from the dinner following the National Nuclear Fuel Cycle Summit, April 5, 2012.

19. Waste Isolation Pilot Plant Land Withdrawal Act, Pub. L. No. 102–579, 106 Stat. 4777 (1992), amended by Pub. L. No. 104–201, 110 Stat. 2422 (1996).

Waste Isolation Pilot Plant Land Withdrawal Amendment Act, enacted as part of National Defense Authorization Act for Fiscal Year 1997, Pub. L. No. 104–201, 110 Stat. 2422, 2851 (1996).

20. WIPP Land Withdrawal Act, P.L. 102–579, 1996.

21. As described on the website of the Energy Communities Alliance, available at http://www.energyca.org/.

22. The City of Carlsbad is a part of Eddy County, which is composed of the City of Carlsbad and the City of Artesia. Following her entry onto the board of the ECA, Lara announced her candidacy for the Democratic Party of southeastern New Mexico for the US Congress in New Mexico's second district. She lost the election. See Simonich (2013).

23. The description of the National Conference of State Legislatures is on the organization's official website: http://www.ncsl.org/research/environment-and-natural-resources/ncsls-high-level-waste-working-group-meeting.aspx (last visited April 2013).

24. Interview with the head of the Task Force, Carlsbad, New Mexico, April 2012.

25. Heaton, J. (2010, July). *The Carlsbad/WIPP History of Transuranic Disposal in Salt.* Presented at the Blue Ribbon Commission on America's Nuclear Future Disposal Subcommittee.

26. The vote concluded with five in favor against three dissenting voices. The dissenting members of the City Board declared that they were concerned about the lack of accountability and lack of transparency of the lobbying firms that are located in Washington, DC. See Davis, S. (2012). "Eddy County Commission Awards $180K Lobbying Bid, Will Split Cost with City of Carlsbad." *Carlsbad Current-Argus*, June 6.

27. Noted in the interview with the head of the Task Force, Carlsbad, New Mexico, April 2012, and noted in the interview with a LANL official, Carlsbad, New Mexico, April 2012.

28. Davis, S. (2012). "DC Lobbyist Hired by Carlsbad, Eddy County Urging DOE to pass WIPP budget." *Carlsbad Current-Argus*, October 28. http://www.currentargus.com/ci_21873901/dc-lobbyist-hired-by-carlsbad-eddy-county-urging (last visited June 12, 2014).

29. See "NO TO WIPP" website: http://www.no2wipp.org/ (last visited May 23, 2014).

30. Hancock, D. (2011, January). *Additional Perspective on the Continuing WIPP Experience*. Presented at the Blue Ribbon Commission on America's Nuclear Future, Carlsbad, NM. http://www.sric.org/nuclear/docs/SRIC%20Statement012711_tables.pdf

31. The SRIC is one of the major environmental watchdogs in northern New Mexico. The organization provides technical and scientific counter-expertise, as well as makes technical and scientific issues public. Hancock joined the SRIC in 1975 after the Three Mile Island accident (McCutcheon 2002, 52).

32. See Heaton (2013).

33. Notes from the official underground visit at WIPP following the National Nuclear Fuel Cycle Summit, Carlsbad, New Mexico, April 5, 2012.

34. Notes from the former mayor's presentation and speech at the WM 2012 Conference, Panel Session 105, *Closing Featured Session US Administration's Blue Ribbon Commission on America's Nuclear Future*, Phoenix, Arizona, March 1, 2012.

35. Interview with the Carlsbad's, Carlsbad, New Mexico, April 2012.

36. The NRC requirements for HLW and SNF retrievability are given in Title 10 of the Code of Federal Regulations (CFR), Part 60 and Part 63.

37. Notes on a presentation to Waste Management 2012 by John Heaton on behalf of the City of Carlsbad, New Mexico, February 2012.

38. In the 1970s and 1980s, many federal studies explored the capacity of salt to confine high-heat generating high-level waste and reached controversial results (Downey 1985).

39. Notes on a presentation by John Heaton at the NCSL Nuclear Workgroup in April 2013.

40. Interview with a senior DOE scientist who contributed to the cost calculation, during the WM, Phoenix, Arizona, March 2012.

41. Notes on a presentation by John Heaton to Waste Management 2012 on behalf of the City of Carlsbad, New Mexico, February 2012.

42. As noted earlier, the map was used in several presentations and in documents produced by National Laboratories and the DOE's Carlsbad field office (e.g., Hansen and Leigh 2011; Heaton 2010).

43. Interview with a senior LANL scientist, Carlsbad, New Mexico, April 2012.

44. Interview with a senior DOE scientist, DOE Carlsbad field office, Carlsbad, New Mexico, April 2012.

45. As it is described on the NEA's official website: http://www.oecd-nea.org/rwm/saltclub/ (last visited April 2013).

46. Scientific controversies have surrounded WIPP since its early development, including debates over brine flow, salt dissolution rates, and container integrity under repository conditions (Downey 1985; Mora 1999; McCutcheon 2002). These controversies have played and continue playing an important role in shaping WIPP's past, present, and future.

47. See Hancock (2011).

48. Field notes from a Task Force meeting, Carlsbad, New Mexico, April 17, 2012.

49. Their views and analysis of actors from the three major sites were made in written and oral comments submitted to the Blue Ribbon Commission by the Washington State Department of Ecology, the South Carolina Department of Health and Environmental Control, the Governor of the State of Idaho, and the Yakama Nation. For instance, see Yakama Nation Environmental Restoration Waste Management Program (2011).

Washington State Department of Ecology. (2011). "Washington State Department of Ecology Comments on 'Blue Ribbon Commission on America's Nuclear Future—Draft Report to the Secretary of Energy," July 29, 2011.

50. These geographies sound distinct, but they are geopolitically entangled, and thus the "commingled" or "unmingled" treatment of different waste types is not straightforward. Even materials categorized as defense high-level vary significantly based on their production sites, generating technologies, and current storage state. These differences have significant temporal, regulatory, spatial, political, and material implications: for instance, the waste at Hanford requires a longer cooling period

before it becomes "manageable" enough for vitrification and subsequent transportation to a different site. See BRC Staff Draft (2011).

51. Interview with a senior SANDIA official, Carlsbad, New Mexico, April 2012.

52. Notes from a Task Force meeting, Carlsbad, New Mexico, April 17, 2012.

53. Nuclear Waste Policy Amendments Act of 1987. 1987. Pub. L. No. 100–203, 101 Stat. 1330.

54. This aim was described by the DOE's Carlsbad field office in a letter to the Environmental Protection Agency as measuring temperatures to confirm heat transfer calculations, monitoring salt movement to validate salt creep calculations, measuring the porosity of reconsolidating salt to validate a salt-reconsolidation model, determining brine and vapor movement for initial information on moisture effects, validating far-field thermal modeling capability by having several interacting alcoves, providing detailed in situ test data for three-dimensional computer code validation and benchmarking, evaluating the chemical effects of various materials placed in proximity to canisters, and associated changes in the near-field chemistry and environment. See US DOE Carlsbad Field Office. (2011, August 11). "Notification of Intent to Begin Salt Disposal Investigations to Mr. Jonathan Edwards, Director of the U.S. Environmental Protection Agency." http://www.epa.gov/radiation/docs/wipp/sdi/sdi_letter_081111.pdf. See US DOE Carlsbad Field Office. (2011). *A Management Proposal for Salt Disposal Investigations with a Field Scale Heater Test at WIPP* (DOE/CBFO-11-3470). Washington, DC: US Department of Energy.

55. See US DOE Carlsbad Field Office (2012).

56. The Sandia field office planned to pursue the project through a series of sequential reviews of the research that it had previously performed at WIPP.

57. Notes from a Task Force meeting, Carlsbad, New Mexico, April 17, 2012.

58. Interview with a senior DOE WIPP scientist, Carlsbad, New Mexico, April 2012.

59. According to a DOE scientist, at the national level, Sandia National Laboratory expected to become the lead laboratory for the new generic repository science program that was to be launched, and SDDI conflicted with their research program.

60. Notes from a Task Force meeting, Carlsbad, New Mexico, April 17, 2012.

61. Interview with a senior DOE WIPP technoscientist, Carlsbad, New Mexico, April 2012.

62. In February 2014, WIPP experienced a radiological accident when a container of transuranic waste ruptured underground, releasing radioactive particles and

contaminating parts of the facility (DOE 2015). The accident forced immediate evacuation and a complete shutdown lasting three years while the DOE investigated the causes and implemented various upgrades. The DOE did not simply restore WIPP to its original operational state but undertook systematic infrastructure modifications: new ventilation systems, additional shaft construction, reconfigured waste handling procedures, and enhanced monitoring systems (for detailed analysis of the WIPP accident, see Ialenti 2021). This "crisis" provided federal agencies with a unique opportunity (Roitman 2014). Since WIPP's reopening, federal agencies have been putting efforts to expand its mission. This has generated and continues to generate political and technoscientific controversies over what constitutes the "right" type of waste fitting the LWA, how waste package volumes should be calculated, and what even constitutes high-level versus transuranic waste. These developments are aligned with the Task Force's objectives. Current efforts concentrate on transforming what WIPP should physically contain through technocratic and administrative procedures without having to revise the political agreement (the LWA) that originally defined the specific responsibilities, temporalities, and values that WIPP was designed to hold together.

Chapter 6

1. "Thoughts and Places" by Abe Van Luik. Carlsbad Caverns, New Mexico. Part 1, the Caverns. http://www.thoughtsandplaces.org/caverns/carlsbad1.html (last visited June 1, 2018).

2. For detailed accounts on the historical transformations that New Mexico has undergone since the late nineteenth century, see Malone and Etulain (1989), Nash (1990, 1999), and DeBuys (2015).

3. WIPP's siting illustrates the complex, often contested, situated, negotiated and nonlinear processes involved in the development of infrastructure projects. After the abandonment of a proposed defense high-level waste repository project in Lyons, Kansas, in 1972, local Carlsbad actors invited federal study of southeastern New Mexico. As the project evolved, President Carter characterized it as an "anomaly" that fell outside emerging national nuclear waste policy framework (Downey 1985). Among other things, WIPP was supposed to be under the regulatory authority of the US Environmental Protection Agency and not the US Nuclear Regulatory Commission. Congressional authorization in 1979 established WIPP as a research and development facility aiming at demonstrating Salado Formation's capacity to confine defense wastes, including defense high-level waste. Throughout its development, significant opposition emerged from northern New Mexico communities, environmental and antinuclear groups, the State of New Mexico, and its congressional representatives. Lawsuits and political negotiations over its siting and mission first resulted in the signature of a Cooperation and Consultation Agreement between the Department of Energy and the State of New Mexico in 1984, which formalized the state's oversight

role and provided funding for the production of counterexpertise (Environmental Evaluation Group) and the 1992 Land Withdrawal Act (amended 1996), which enacted WIPP's mission into law as limited to military transuranic waste.

4. From the comments by Reverend Rogers, a Task Force member, in front of the Blue Ribbon Commission on America's Nuclear Futures, in Carlsbad, New Mexico. Official transcript of the meeting, January 27, 2011.

5. Social studies of infrastructures have long examined how decisions about siting, design, and operation embed values and visions of order while generating promises of progress and threats of disruption (Star 1999; Bowker and Star 2000; Barry 2006, 2013, 2020; Edwards et al. 2009). This scholarship reveals infrastructures as sites of contestation where different futures, risk assessments, and definitions of public good collide—showing how technical arrangements simultaneously enact social and political ones (Carse 2012; Harvey and Knox 2015; Anand et al. 2018). Rather than neutral technical objects, infrastructures actively shape social relations, distribute benefits and burdens unevenly, and create evolving forms of dependence and vulnerability (Hetherington 2018; Harrington 2023).

6. Scholarship on infrastructure siting has been marked by significant analytical tensions. Several authors have characterized opposition movements in negative terms, framing their members as motivated by selfishness, irrationality, or narrow self-interest rather than legitimate concerns about risk, equity, or democratic participation (see, e.g., Mazmanian and Morell 1990; Kraft and Clary 1991; Dear 1992; Freudenburg and Pastor 1992). This approach often employs the NIMBY (Not In My Back Yard) framework, which often dismisses local opposition as inherently obstructionist. However, subsequent scholarship has critiqued this negative characterization, arguing that the NIMBY label obscures the substantive concerns that motivate opposition movements and fails to account for the complex ways communities engage with infrastructure proposals. These studies demonstrate that opposition often stems from legitimate concerns about environmental justice, democratic process, and the unequal distribution of risks and benefits rather than "selfishness" (see, e.g., Wolsink 1994, 2006; Gordon and Jasper 1996; Wexler 1996; Trom 1999; Burningham 2000; Devine-Wright 2005; Sébastien 2013).

7. A substantial body of risk literature has attempted to calculate monetary compensations for the negative "externalities" generated by infrastructure siting. These studies develop various models for quantifying and pricing environmental and social costs, from contingent valuation methods that survey willingness to pay to avoid risks to cost-benefit analyses that assign monetary values to environmental damages and health impacts. This approach renders complex harms commensurable through monetary valuation. See, for instance, O'Hare (1977), Portney (1985), Mitchell and Carson (1986), Kunreuther et al. (1987), O'Hara (1992), Been (1993), O'Hare and Sanderson (1993), Kunreuther (1995), and Groothuis et al. (1998).

8. Comment by a Carlsbad resident supporting the city's nuclear endeavor before the Blue Ribbon Commission on America's Nuclear Futures, in Carlsbad, New Mexico. Official transcript of the meeting, January 27, 2011.

9. A Task Force member's comment before the Blue Ribbon Commission as a local resident in Carlsbad, New Mexico. Official transcript of the hearing, January 27, 2011.

10. Interview with a senior LANL scientist, Carlsbad, New Mexico, April 2012.

11. Nuclear Waste Policy Act Amendments Act of 1987, Pub.L. No. 100–203, 101, Stat. 1330 (1986).

12. For more details on the project, see Erickson et al. (1994), Sachs (1996), Gowda (1998), Gowda and Easterling (2000), Rogers and Kingsley (2004), and Houston (2013).

13. Their idea was to store spent fuel rods in 16-foot-high, air-cooled, dry storage casks constructed from steel and concrete. Each cask would have been designed to contain about ten tons of spent fuel rods and would sit upon a concrete bed above ground in the desert. Notes from an interview with the Chairman of the Board and the CEO of the Private Fuel Storage LLC, Phoenix, Arizona, March 2012.

14. Interview with the Chairman of the Board and the CEO of the Private Fuel Storage LLC, Phoenix, Arizona, March 2012.

15. I received the invitation from my key interlocutor, who facilitated my attendance at the meeting.

16. Notes from the meeting organized for Senator Murkowski, Pecos River Conference Centre, Carlsbad, New Mexico, April 10, 2012.

17. Maine Yankee Atomic Power Company, Connecticut Yankee Atomic Power Company, & Yankee Atomic Electric Company. (2013, February 6). "In Win for Ratepayers Federal Government Pays Companies Nearly $160 Million." http://www.connyankee.com/_pdf/Phase1DOELitigation02062013.pdf (last visited January 2015).

18. Since November 2006, *Electricité de France* has owned 49.9 percent of the nuclear subsidiary of Constellation Energy. See (US Securities and Exchange Commission 2015). http://www.sec.gov/Archives/edgar/data/9466/000104746912001863/a2207433z10-k.htm (last visited January 5, 2015).

19. US Securities and Exchange Commission. (2015). *Annual Report for the Fiscal Year Ended December 31, 2011—Constellation Energy Group, INC.* (No. 10-K, 1–12869). Washington, DC: U.S. Securities and Exchange Commission. http://www.sec.gov/Archives/edgar/data/9466/000104746912001863/a2207433z10-k.htm (last visited January 5, 2015).

20. Principles for Safeguarding Nuclear Waste at Reactors. https://www.clearwater.org/wp-content/uploads/2019/10/Principles-for-Safeguarding-Nuclear-Waste-at-Reactors-7.12.18.pdf (last visited January 2022).

21. S.2009 109th Congress, 1st Session. See (Nevada Commission on Nuclear Projects 2006).

Principles for Safeguarding Nuclear Waste at Reactors (2010). http://www.psr.org/nuclear bailout/resources/principles-for-safeguarding.pdf.

22. During this time, GNEP and the Office of Nuclear of Energy of the DOE held several Programmatic Environmental Impact Statement Hearings at different sites in New Mexico. See Department of Energy, Office of Nuclear Energy (2007, 2008a, 2008b, 2008c, 2008d).

23. Notes from an interview with the head of the Task Force, Carlsbad, New Mexico, April 2012.

24. Interview with the head of the Task Force, Carlsbad, New Mexico, April 2012.

25. See AREVA (2012).

26. As of 2018, AREVA no longer exists. It has been reorganized to concentrate on the back end of the nuclear fuel cycle and has been renamed as ORANO.

27. Interview with a senior AREVA official, Phoenix, Arizona, March 2012.

28. Since the NRC's 2014 ruling, multiple private companies have been pursuing private centralized interim storage projects in Texas and New Mexico, including Holtec International working with the Eddy-Lea Energy Alliance. These proposals have been licensed by the Nuclear Regulatory Commission for an initial period of forty years but have been increasingly facing opposition from state governments, environmental groups, Indigenous communities, and, perhaps unexpectedly, oil and gas producers concerned about impacts on their economic activities. Political and legal battles have ensued at both state and federal levels, with state governments challenging the projects through legislation and federal court proceedings. These facilities have been approved through standard Environmental Impact Statement procedures, as conducted by the Nuclear Regulatory Commission, that provide no meaningful avenue for extended negotiation and political agreement. This shift represents what I have elsewhere termed the "end of an engagement with deep time" (Saraç-Lesavre 2021b).

29. Interview with a senior LANL scientist, Carlsbad, New Mexico, April 2012.

Bibliography

Ackerman, Frank, and Lisa Heinzerling. 2002. "Pricing the Priceless: Cost-Benefit Analysis of Environmental Protection." *University of Pennsylvania Law Review* 150 (5): 1553–1584.

Adam, Barbara, and Chris Groves. 2007. *Future Matters: Action, Knowledge, Ethics.* Brill Academic.

Akrich, Madeleine, Michel Callon, and Bruno Latour. 2006. *Sociologie de la traduction: textes fondateurs*. Paris: Presses des Mines.

Alexander, Catherine, and Joshua Reno. 2012. *Economies of Recycling: The Global Transformation of Materials, Values and Social Relations*. Bloomsbury Publishing.

Alexander, Catherine, and Patrick O'Hare. 2023. "Waste and Its Disguises: Technologies of (Un)Knowing." *Ethnos* 88 (3): 419–443.

Alexis-Martin, Becky, and Thom Davies. 2017. "Towards Nuclear Geography: Zones, Bodies, and Communities." *Geography Compass* 11 (9): e12325.

Alley, William M., and Rosemarie Alley. 2013. *Too Hot to Touch: The Problem of High-Level Nuclear Waste*. Cambridge University Press.

Anand, Nikhil, Akhil Gupta, and Hannah Appel. 2018. *The Promise of Infrastructure.* Duke University Press.

Anderson, Benedict R. O'G. 2006. *Imagined Communities: Reflections on the Origin and Spread of Nationalism*. Rev. ed. Verso.

Anderson, Gary M. 1991. "The Fiscal Significance of User Charges and Earmarked Taxes: A Survey." In *Charging for Government: User Charges and Earmarked Taxes in Principle and Practice*, edited by Richard E. Wagner. Taylor & Francis.

Antal, Ariane Berthoin, Michael Hutter, and David Stark. 2015. *Moments of Valuation: Exploring Sites of Dissonance*. Oxford University Press.

AREVA. 2012. "Areva Led Team Selected by Eddy-Lea Energy Alliance LLC to Develop Interim Consolidated Storage Facility." Press release. http://us.areva.com/EN/home-2016/areva-eddy-lea-unf-storage-facility.html.

Arnaboldi, Michela, and Tommaso Palermo. 2011. "Translating Ambiguous Reforms: Doing Better next Time?" *Management Accounting Research* 22 (1): 6–15.

Arrow, Kenneth J., and Robert C. Lind. 1970. "Uncertainty and the Evaluation of Public Investment Decision." *American Economic Review* 60: 364–378.

Atomic Safety and Licensing Board, U.S. Nuclear Regulatory Commission. 2010a. "Memorandum and Order. In the Matter of U.S. Department of Energy (High-Level Waste Repository)." Docket No. 63-001-HLW ASLBP No. 09-892-HLW-CAB04. NRC, June 29, 2010.

Atomic Safety and Licensing Board, U.S. Nuclear Regulatory Commission. 2010b. "U.S. Department of Energy's Motion to Withdraw, In the Matter of U.S. Department of Energy (High-Level Waste Repository)." Docket No. 63-001-HLW ASLBP No. 09-892-HLW-CAB04. NRC.

Ballestero, Andrea. 2015. "The Ethics of a Formula: Calculating a Financial–Humanitarian Price for Water." *American Ethnologist* 42 (2): 262–278.

Ballestero, Andrea. 2019. *A Future History of Water*. Duke University Press.

Ballestero, Andrea, Andrea Muehlebach, and Gloria Pérez-Rivera. 2023. "What Is a Financial Frontier?" *Journal of Cultural Economy* 16 (3): 311–322.

Banoub, Daniel, and Stephen J. Martin. 2020. "Storing Value: The Infrastructural Ecologies of Commodity Storage." *Environment and Planning D: Society and Space* 38 (6): 1101–1119.

Barad, Karen. 2019. "After the End of the World: Entangled Nuclear Colonialisms, Matters of Force, and the Material Force of Justice." *Theory & Event* 22 (3): 524–550.

Barbier, Rémi. 2005. "Le Financement de l'élimination Des Déchets Municipaux." *Annuaire Des Collectivités Locales* 25 (1): 87–99.

Barretta, Antonio, and Cristiano Busco. 2011. "Technologies of Government in Public Sector's Networks: In Search of Cooperation through Management Control Innovations." *Management Accounting Research* 22 (4): 211–219.

Barry, Andrew. 2002. "The Anti-Political Economy." *Economy and Society* 31 (2): 268–284.

Barry, Andrew. 2006. "Technological Zones." *European Journal of Social Theory* 9 (2): 239–253.

Barry, Andrew. 2013. *Material Politics: Disputes along the Pipeline*. John Wiley & Sons.

Barry, Andrew. 2020. "The Material Politics of Infrastructure." In *TechnoScienceSociety*, edited by Sabine Maasen, Sascha Dickel, and Christoph Schneider, 30: 91–109. Sociology of the Sciences Yearbook. Springer International Publishing.

Baumeister, Roy F., and Kathleen D. Vohs. 2003. "Self-Regulation and the Executive Function of the Self." *Handbook of Self and Identity* 1: 197–217.

Baumol, William F. 1968. "On the Social Rate of Discount." In *Classic Papers in Natural Resource Economics*, 37–53. Palgrave Macmillan UK.

Bazelon, Coleman, and Kent Smetters. 2001. "Discounting in the Long-term." *Loyola of Los Angeles Law Review* 35: 277.

Beckert, Jens. 2013. "Imagined Futures: Fictional Expectations in the Economy." *Theory and Society* 42 (3): 219–240.

Beckert, Jens. 2016. *Imagined Futures: Fictional Expectations and Capitalist Dynamics*. Harvard University Press.

Beckert, Jens. 2024. *How We Sold Our Future: The Failure to Fight Climate Change*. John Wiley & Sons.

Beckert, Jens, and Lisa Suckert. 2021. "The Future as a Social Fact. The Analysis of Perceptions of the Future in Sociology." *Poetics* 84: 101499.

Beckert, Jens, and Patrik Aspers, eds. 2011. *The Worth of Goods: Valuation and Pricing in the Economy*. Oxford University Press.

Beckert, Jens, and Richard Bronk. 2018. *Uncertain Futures: Imaginaries, Narratives, and Calculation in the Economy*. Oxford University Press.

Been, Vicki. 1993. "Compensated Siting Proposals: Is It Time to Pay Attention?" *Fordham Urban Law Journal* 21, no. 3: 787.

Bell, Marissa. 2023. "The Epistemic Tensions of Nuclear Waste Siting in a Nuclear Landscape." *Environment and Planning E: Nature and Space* 6 (2): 841–862.

Bell, Marissa. 2024. "'We Are a Nuclear Community': The Ethical, Political Economic, and Social Relations of Canadian Nuclear Waste Siting." *Antipode* 57 (4): 1299–1319.

Benford, G. 2000. *Deep Time: How Humanity Communicates Across Millennia*. Harper Collins.

Bergmans, Anne, Göran Sundqvist, Drago Kos, and Peter Simmons. 2015. "The Participatory Turn in Radioactive Waste Management: Deliberation and the Social–Technical Divide." *Journal of Risk Research* 18 (3): 347–363.

Berman, Elizabeth P. 2022. *Thinking like an Economist*. Princeton University Press.

Besnier, Niko, and Susana Narotzky. 2018. "Epilogue. Indeterminacy: Between Worth and Worthlessness." In *Indeterminacy: Waste, Value, and the Imagination*, edited by Catherine Alexander and Andrew Sanchez. Bergahn.

Bezes, Philippe, and Alexandre Siné, eds. 2011. *Gouverner (Par) Les Finances Publiques*. Presses de Sciences Po.

Bijker, Wiebe E. 1997. *Of Bicycles, Bakelites, and Bulbs: Toward a Theory of Sociotechnical Change*. MIT Press.

Bijker, Wiebe E., Thomas P. Hughes, and Trevor J. Pinch. 1987. *The Social Construction of Technological Systems: New Directions in the Sociology and History of Technology*. MIT Press.

Binkley, S. 2009. "Governmentality, Temporality and Practice: From the Individualization of Risk to the 'contradictory Movements of the Soul'." *Time & Society* 18 (1): 86–105.

Birch, Kean, and Fabian Muniesa. 2020. *Assetization: Turning Things into Assets in Technoscientific Capitalism*. MIT Press.

Blowers, Andrew, and Göran Sundqvist. 2010. "Radioactive Waste Management–Technocratic Dominance in an Age of Participation." *Journal of Integrative Environmental Sciences* 7 (3): 149–155.

Blue Ribbon Commission on America's Nuclear Future. 2012. "Report to the Secretary of Energy." Blue Ribbon Commission on America's Nuclear Future.

Bogener, Stephen. 2008. "Land, Speculation, and Manipulation on the Pecos." *Great Plains Quarterly* 28 (3): 209–229.

Borup, M., N. Brown, K. Konrad, and H. Van Lente. 2006. "The Sociology of Expectations in Science and Technology." *Technology Analysis & Strategic Management* 18 (3): 285–298.

Boston, Jonathan. 2017. *Governing for the Future: Designing Democratic Institutions for a Better Tomorrow*. Emerald Publishing.

Boston, Jonathan. 2021a. "Assessing the Options for Combatting Democratic Myopia and Safeguarding Long-Term Interests." *Futures* 125: 102668.

Boston, Jonathan. 2021b. "Protecting Long-Term Interests: The Role of Institutions as Commitment Devices." In *Giving Future Generations a Voice*, 86–107. Edward Elgar.

Bourdieu, Pierre. 1990. "Time Perspectives of the Kabyle." In *The Sociology of Time*, edited by John Hassard, 219–37. Springer.

Bowker, Geoffrey C., and Susan L. Star. 2000. *Sorting Things Out: Classification and Its Consequences*. MIT Press.

Brady, Michelle. 2014. "Ethnographies of Neoliberal Governmentalities: From the Neoliberal Apparatus to Neoliberalism and Governmental Assemblages." *Foucault Studies* 18 (2014): 11–33.

Brown, Kate. 2016. "The Last Sink: The Human Body as the Ultimate Radioactive Storage Site." *RCC Perspectives* 1: 41–48.

Brown, Kate. 2019. *Manual for Survival: A Chernobyl Guide to the Future*. Penguin.

Brown, Wendy. 2006. "American Nightmare: Neoliberalism, Neoconservatism, and De-Democratization." *Political Theory* 34 (6): 690–714.

Browne, Katherine E. 2008. "Economics and Morality: Introduction." In *Economics and Morality: Anthropological Approaches*, edited by Katherine E. Browne and B. Lynne Milgram, 1–40. Rowman & Littlefield. https://doi.org/10.5771/9780759112261.

BRC Staff Draft. 2011. "Background Paper on Commingling of Defense and Commercial Waste." November 17, 2011.

Bryant, Rebecca, and Daniel M. Knight. 2019. *The Anthropology of the Future*. Cambridge University Press.

Bubandt, Nils, and Anna Tsing. 2018. "Feral Dynamics of Post-Industrial Ruin: An Introduction." *Journal of Ethnobiology* 38 (1): 1.

Buchanan, James M., and Richard A. Musgrave. 1999. *Public Finance and Public Choice: Two Contrasting Visions of the State*. MIT Press.

Burningham, Kate. 2000. "Using the Language of NIMBY: A Topic for Research, Not an Activity for Researchers." *Local Environment* 5 (1): 55–67.

Butt, Daniel. 2006. "Nations, Overlapping Generations, and Historic Injustice." *American Philosophical Quarterly* 43 (4): 357–367.

Callon, Michel. 1986. "Some Elements of a Sociology of Translation: Domestication of the Scallops and the Fishermen of St. Brieuc Bay." *Power, Action, and Belief: A New Sociology of Knowledge* 32: 196–223.

Callon, Michel. 1998. "An Essay on Framing and Overflowing." In *The Laws of the Markets*, edited by Michel Callon, 244–269. Blackwell.

Callon, Michel. 2009. "Postface. La formulation marchande des biens." In *Evaluer et valoriser: une sociologie économique de la mesure*, edited by François Vatin, 247–269. Presses Universitaires du Mirail.

Callon, Michel, and Fabian Muniesa. 2005. "Peripheral Vision: Economic Markets as Calculative Collective Devices." *Organization Studies* 26 (8): 1229–1250.

Callon, Michel, Cécile Méadel, and Vololona Rabeharisoa. 2002. "The Economy of Qualities." *Economy and Society* 31 (2): 194–217.

Callon, Michel, Yannick Barthe, and Pierre Lascoumes. 2009. *Acting in an Uncertain World: An Essay on Technical Democracy*. MIT Press.

Carse, Ashley. 2012. "Nature as Infrastructure: Making and Managing the Panama Canal Watershed." *Social Studies of Science* 42 (4): 539–563.

Chakrabarty, Dipesh. 2018. "Anthropocene Time." *History and Theory* 57 (1): 5–32.

Chakrabarty, Dipesh. 2021. *The Climate of History in a Planetary Age*. University of Chicago Press.

Chrulew, Matthew. 2017. "Freezing the Ark: The Cryopolitics of Endangered Species Preservation." In *Cryopolitics: Frozen Life in a Melting World*, 283–300. MIT Press.

Chua, Charmaine, Martin Danyluk, Deborah Cowen, and Laleh Khalili. 2018. "Introduction: Turbulent Circulation: Building a Critical Engagement with Logistics." *Environment and Planning D: Society and Space* 36 (4): 617–629.

City of Carlsbad. 2012. *Greater Carlsbad Comprehensive Plan: Strategy 2030*. City of Carlsbad.

Claiborne, H. C. 1977. "The United States Program for the Safety Assessment of Geological Disposal of Commercial Radioactive Wastes." In *Workshop on Risk Analysis and Geological Modelling*. Oak Ridge National Laboratory.

Claiborne, H. C., and Ferruccio Gera. 1974. "Potential Containment Failure Mechanisms and Their Consequences at a Radioactive Waste Repository in Bedded Salt in New Mexico." Oak Ridge National Laboratory.

Clark, Nigel, and Bronislaw Szerszynski. 2020. *Planetary Social Thought: The Anthropocene Challenge to the Social Sciences*. Polity Press.

Coe, Neil M. 2020. "Logistical Geographies." *Geography Compass* 14 (10): e12506.

Coleman, Rebecca. 2017. "A Sensory Sociology of the Future: Affect, Hope and Inventive Methodologies." *The Sociological Review* 65 (3): 525–543.

Congressional Research Service. 2012. "Closing Yucca Mountain: Litigation Associated with Attempts to Abandon the Planned Nuclear Waste Repository." 7-5700/R41675. Congressional Research Service.

Cowan, George A. 1977. "Migration Paths for Oklo Reactor Products and Applications to the Problem of Geological Storage of Nuclear Wastes." Report No. LA-UR-77-2787. Los Alamos National Laboratory.

Cram, Shannon. 2015. "Becoming Jane: The Making and Unmaking of Hanford's Nuclear Body." *Environment and Planning D: Society and Space* 33 (5): 796–812.

Cram, Shannon. 2016. "Wild and Scenic Wasteland: Conservation Politics in the Nuclear Wilderness." *Environmental Humanities* 7 (1): 89–105.

Cruz, Luis. 2023. "Environmental Justice Considerations in Siting Spent Nuclear Fuel Disposal." *The George Washington Law Review* 91: 499.

Danyluk, Martin. 2018. "Capital's Logistical Fix: Accumulation, Globalization, and the Survival of Capitalism." *Environment and Planning D: Society and Space* 36 (4): 630–647.

Davies, Thom, and Abel Polese. 2015. "Informality and Survival in Ukraine's Nuclear Landscape: Living with the Risks of Chernobyl." *Journal of Eurasian Studies* 6 (1): 34–45.

Davies, Will. 2017. "Moral Economies of the Future-the Utopian Impulse of Sustainable Prosperity." CUSP Working Paper Series, no. 5. https://research.gold.ac.uk/id/eprint/23504/

Davis, Stella. 2012. "New Mexico Auditor's Office Taking a Closer Look at Eddy-Lea Energy Alliance." *Carlsbad Current-Argus*, June 10, 2012.

Dear, Michael. 1992. "Understanding and Overcoming the NIMBY Syndrome." *Journal of the American Planning Association* 58 (3): 288–300.

DeBuys, William. 2015. *Enchantment and Exploitation: The Life and Hard Times of a New Mexico Mountain Range*. University of New Mexico Press.

Deringer, William. 2017. "Pricing the Future in the Seventeenth Century: Calculating Technologies in Competition." *Technology and Culture* 58 (2): 506–528.

Deringer, William. 2018. *Calculated Values: Finance, Politics, and the Quantitative Age*. Harvard University Press.

Deringer, William. 2024. "Mr. Aecroid's Tables: Economic Calculations and Social Customs in the Early Modern Countryside." *Journal of Modern History* 96 (1): 1–46.

Devine-Wright, Patrick. 2005. "Beyond NIMBYism: Towards an Integrated Framework for Understanding Public Perceptions of Wind Energy." *Wind Energy* 8 (2): 125–139.

Dewey, John. 1939. "Theory of Valuation." *International Encyclopedia of Unified Science* 2, no. 4. University of Chicago Press.

Dianderas, Eben R. 2024. "Volumes: The Politics of Calculation in Contemporary Peruvian Amazonia." *Cultural Anthropology* 39 (1): 64–90.

Doganova, Liliana. 2018a. "Discounting and the Making of the Future." In *Uncertain Futures*, edited by Jens Beckert and Richard Bronk, 278–297. Oxford University Press.

Doganova, Liliana. 2018b. "Discounting the Future: A Political Technology." *Economic Sociology_the European Electronic Newsletter* 19 (2): 4–9.

Doganova, Liliana. 2024. *Discounting the Future: The Ascendancy of a Political Technology*. Zone Books.

Downey, Gary L. 1985. "Federalism and Nuclear Waste Disposal: The Struggle over Shared Decision Making." *Journal of Policy Analysis and Management* 5 (1): 73–99.

Edwards, Paul N., Geoffrey C. Bowker, Steven J. Jackson, and Robin Williams. 2009. "Introduction: An Agenda for Infrastructure Studies." *Journal of the Association for Information Systems* 10 (5): 6.

Endres, Danielle. 2009. "From Wasteland to Waste Site: The Role of Discourse in Nuclear Power's Environmental Injustices." *Local Environment* 14 (10): 917–937.

Endres, Danielle. 2012. "Sacred Land or National Sacrifice Zone: The Role of Values in the Yucca Mountain Participation Process." *Environmental Communication: A Journal of Nature and Culture* 6 (3): 328–345.

Erickson, Jon D., Duane Chapman, and Ronald E. Johnny. 1994. "Monitored Retrievable Storage of Spent Nuclear Fuel in Indian Country: Liability, Sovereignty, and Socioeconomics." *American Indian Law Review* 19: 73–103.

Esposito, Elena. 2018. "The Time of Money in Finance and US Society." *Finance and Society* 4 (1): 15–25.

Evans-Pritchard, Edward Evan. 1939. "Nuer Time-Reckoning." *Africa* 12 (2): 189–216.

Ewing, Rodney C. 1999. "Less Geology in the Geological Disposal of Nuclear Waste." *Science* 286 (5439): 415–417. https://doi.org/10.1126/science.286.5439.415.

Ezzamel, Mahmoud. 2009. "Order and Accounting as a Performative Ritual: Evidence from Ancient Egypt." *Accounting, Organizations and Society* 34 (3–4): 348–380.

Fassin, Didier. 2008. "Beyond Good and Evil? Questioning the Anthropological Discomfort with Morals." *Anthropological Theory* 8 (4): 333–344.

Fassin, Didier, and Jean-Sébastien Eideliman. 2014. *Économies Morales Contemporaines*. La découverte.

Faustmann, Martin. 1849. "Berechnung Des Werthes, Welchen Waldboden, Sowie Noch Nicht Haubare Holzbestande Fur Die Waldwirthschaft Besitzen [Calculation of the Value Which Forest Land and Immature Stands Possess for Forestry]." *Allgemeine Fotst-Und Jagd-Zeitung* 25: 441–455.

Ferry, Elizabeth E., and Mandana E. Limbert. 2008. *Timely Assets: The Politics of Resources and Their Temporalities*. School for Advanced Research Press.

Fields, Gary. 2011. "Enclosure Landscapes: Historical Reflections on Palestinian Geography." *Historical Geography* 39: 182–207.

Flynn, J., ed. 1995. *One Hundred Centuries of Solitude: Redirecting America's High-Level Nuclear Waste Policy*. Westview Press.

Foucault, Michel. 2010. *The Birth of Biopolitics: Lectures at the Collège de France, 1978—1979*. Picador.

Fourcade, Marion. 2011. "Cents and Sensibility: Economic Valuation and the Nature of 'Nature.'" *American Journal of Sociology* 116 (6): 1721–1777.

Frankel, Christian, José Ossandón, and Trine Pallesen. 2019. "The Organization of Markets for Collective Concerns and Their Failures." *Economy and Society* 48 (2): 153–174.

Frederick, Shane, George Loewenstein, and Ted O'Donoghue. 2002. "Time Discounting and Time Preference: A Critical Review." *Journal of Economic Literature* 40: 351–401.

Fressoz, Jean-Baptiste. 2012. *L'Apocalypse Joyeuse. Une Histoire Du Risque Technologique*. Le Seuil.

Fressoz, Jean-Baptiste. 2013. "Payer Pour Polluer—L'industrie Chimique et La Compensation Des Dommages Environnementaux, 1800–1850." *Histoire & Mesure* 28 (1): 145–186.

Freudenburg, William R., and Susan K. Pastor. 1992. "NIMBYs and LULUs: Stalking the Syndromes." *Journal of Social Issues* 48 (4): 39–61.

Fridman, Daniel. 2010. "From Rats to Riches: Game Playing and the Production of the Capitalist Self." *Qualitative Sociology* 33 (4): 423–446.

Fridman, Daniel. 2014. "Resisting the Lure of the Paycheck: Freedom and Dependence in Financial Self-Help." *Foucault Studies* 18: 90–112.

Gailliot, Matthew T., and Roy F. Baumeister. 2007. "The Physiology of Willpower: Linking Blood Glucose to Self-Control." *Personality and Social Psychology Review* 11 (4): 303–327.

Gaines, Sanford E. 1991. "The Polluter-Pays Principle: From Economic Equity to Environmental Ethos." *Texas International Law Journal* 26: 463.

Galison, Peter. 1987. *How Experiments End*. University of Chicago Press.

Galison, Peter. 2000. "Einstein's Clocks: The Place of Time." *Critical Inquiry* 26 (2): 355–389.

Galison, Peter. 2004. *Einstein's Clocks, Poincaré's Maps: Empires of Time*. Norton.

Galison, Peter. 2014. "The Future of Scenarios: State Science Fiction." In *The Subject of Rosi Braidotti: Politics and Concepts*, 38–46. Bloomsbury Publishing.

Galison, Peter, and J. Kruse. 2012. "Waste-Wilderness: A Conversation with Peter L. Galison." *Discard Studies* (blog). 2012.

Galison, Peter, and Robb Moss, dirs. 2015. *Containment*. Bullfrog Films.

Gell, Alfred. 2021. *The Anthropology of Time: Cultural Constructions of Temporal Maps and Images*. Routledge.

Genay, Lucie. 2019. *Land of Nuclear Enchantment: A New Mexican History of the Nuclear Weapons Industry*. University of New Mexico Press.

Gollier, Christian. 2002. "Discounting an Uncertain Future." *Journal of Public Economics* 85 (2): 149–166.

Gollier, Christian, and Martin L. Weitzman. 2010. "How Should the Distant Future Be Discounted When Discount Rates Are Uncertain?" *Economics Letters* 107 (3): 350–353.

Gómez, Myrriah. 2022. *Nuclear Nuevo México: Colonialism and the Effects of the Nuclear Industrial Complex on Nuevomexicanos*. University of Arizona Press.

Goodin, Robert E. 1978. "Uncertainty as an Excuse for Cheating Our Children: The Case of Nuclear Wastes." *Policy Sciences* 10 (1): 25–43.

Gordillo, Gastón R. 2014. *Rubble: The Afterlife of Destruction*. Duke University Press.

Gordon, Cynthia, and James M. Jasper. 1996. "Overcoming the 'NIMBY' Label: Rhetorical and Organizational Links for Local Protestors." *Research in Social Movements, Conflicts and Change* 19: 153–175.

Gosseries, Axel. 2008. "On Future Generations' Future Rights." *Journal of Political Philosophy* 16 (4): 446–474.

Gosseries, Axel. 2023. *What Is Intergenerational Justice?* John Wiley & Sons.

Gowda, M. V. R. 1998. "Nuclear Waste and Native America: The MRS Siting Exercise." *Risk: Health, Safety & Environment* 9: 229.

Gowda, M. V. R., and Doug Easterling. 2000. "Voluntary Siting and Equity: The MRS Facility Experience in Native America." *Risk Analysis* 20 (6): 917–930.

Graham, John D. 2007. "Valuing the Future: OMB's Refined Position." *University of Chicago Law Review* 74 (Winter): 51.

Groothuis, Peter A., George Van Houtven, and John C. Whitehead. 1998. "Using Contingent Valuation to Measure the Compensation Required to Gain Community Acceptance of a LULU: The Case of a Hazardous Waste Disposal Facility." *Public Finance Review* 26 (3): 231–249.

Grove, Benjamin. 2003. "Reid Panel Cuts Yucca Budget." *Las Vegas Sun*, July 16.

Gunningham, Neil. 2009. "Environment Law, Regulation and Governance: Shifting Architectures." *Journal of Environmental Law* 21 (2): 179–212.

Gupta, Kuhika, Hank Jenkins-Smith, Joseph Ripberger, Andrew Fox, and Will Livingston. 2024. "Consent-Based Siting of Spent Fuel Facilities: Evidence from Nationwide U.S. Surveys." *Nuclear Technology* 210 (9): 1754–1766.

Gusterson, Hugh. 1998. *Nuclear Rites: A Weapons Laboratory at the End of the Cold War*. University of California Press.

Gusterson, Hugh. 2000. "How Not to Construct a Radioactive Waste Incinerator." *Science, Technology & Human Values* 25 (3): 332–351.

Guyer, Jane I. 2007. "Prophecy and the near Future: Thoughts on Macroeconomic, Evangelical, and Punctuated Time." *American Ethnologist* 34 (3): 409–421.

Guyer, Jane I. 2014. "Gains and Losses in the Margins of Time: From West and Equatorial History to Present-Day South Africa, and Back." *Africa* 84 (1): 146–150.

Hache, Emilie. 2007a. "La responsabilité, une technique de gouvernementalité néolibérale?" *Raisons Politiques* 28 (4): 49. https://doi.org/10.3917/rai.028.0049.

Hache, Emilie. 2007b. "Néolibéralisme et responsabilité." *Raisons Politiques* 28 (4): 5.

Hale, Thomas. 2024. *Long Problems: Climate Change and the Challenge of Governing Across Time*. Princeton University Press.

Hamblin, Jacob D. 2008. *Poison in the Well: Radioactive Waste in the Oceans at the Dawn of the Nuclear Age*. Rutgers University Press.

Hancock, Don. 2011. "Additional Perspective on the Continuing WIPP Experience." Presented at the Blue Ribbon Commission on America's Nuclear Future, Carlsbad, New Mexico, January. http://www.sric.org/nuclear/docs/SRIC%20Statement012711_tables.pdf

Hansen, Frank D., and Christi D. Leigh. 2011. "Salt Disposal of Heat-Generating Nuclear Waste." SAND2011-0161. Sandia National Laboratories.

Hanson, Randel D. 2001. "An Experiment in (Toxic) Indian Capitalism?: The Skull Valley Goshutes, New Capitalism, and Nuclear Waste." *PoLAR: Political and Legal Anthropology Review* 24 (2): 25–38.

Hanusch, Frederic. 2023. *The Politics of Deep Time*. Cambridge University Press.

Hanusch, Frederic, and Frank Biermann. 2020. "Deep-Time Organizations: Learning Institutional Longevity from History." *The Anthropocene Review* 7 (1): 19–41.

Haraway, Donna J. 2016. *Staying with the Trouble: Making Kin in the Chthulucene*. Duke University Press.

Harrington, Cameron. 2023. "The Eternal Return: Imagining Security Futures at the Doomsday Vault." *Environment and Planning E: Nature and Space* 6 (4): 2614–2635.

Hartog, François. 2003. *Régimes d'historicité: Présentisme et Expériences Du Temps*. Seuil.

Harvey, Penny, and Hannah Knox. 2015. *Roads: An Anthropology of Infrastructure and Expertise*. Cornell University Press.

Hassard, John. 1990. "Introduction: The Sociological Study of Time." In *The Sociology of Time*, 1–18. Palgrave Macmillan UK.

Hassard, John. 1990. *The Sociology of Time*. Springer.

Hawkins, Gay, Emily Potter, and Kane Race. 2015. *Plastic Water: The Social and Material Life of Bottled Water*. MIT Press.

Heaton, John. 2010. "The Carlsbad/WIPP History of Transuranic Disposal in Salt." Presented at the Blue Ribbon Commission on America's Nuclear Future Disposal Subcommittee, July 7.

Heaton, John. 2013, April. "WIPP: A Story of Consent." Presented at the NCSL Nuclear Workgroup.

Hecht, Gabrielle. 2006. "Negotiating Global Nuclearities: Apartheid, Decolonization, and the Cold War in the Making of the IAEA." *Osiris*, 2nd Series, 21 (January): 25–48.

Hecht, Gabrielle, ed. 2011. *Entangled Geographies: Empire and Technopolitics in the Global Cold War*. Inside Technology. MIT Press.

Hecht, Gabrielle. 2012. *Being Nuclear: Africans and the Global Uranium Trade*. MIT Press.

Heinzerling, Lisa. 1999. "Discounting Our Future." *Land & Water Law Review* 34: 39.

Heinzerling, Lisa. 2001. "The Temporal Dimension in Environmental Law." *Environmental Law Reporter* 31 (9): 11055–11072.

Heinzerling, Lisa. 2006. "The Accidental Environmentalist: Judge Posner on Catastrophic Thinking." *Georgetown Law Journal* 94: 833–857.

Helgesson, Claes-Fredrik, and Fabian Muniesa. 2013. "For What It's Worth: An Introduction to Valuation Studies." *Valuation Studies* 1 (1): 1–10.

Helman, Christopher. 2012. "The Town That Wants America's Worst Atomic Waste." *Forbes Magazine*, January 25, 2012.

Herstein, Ori J. 2008. "The Identity and (Legal) Rights of Future Generations." *The George Washington Law Review* 77: 1173.

Hetherington, Kevin. 2004. "Secondhandedness: Consumption, Disposal, and Absent Presence." *Environment and Planning D* 22 (1): 157–174.

Hetherington, Kregg. 2018. *Infrastructure, Environment, and Life in the Anthropocene*. Duke University Press.

Hezir, Joseph. 2011. "Budget and Financial Management Improvements to the Nuclear Waste Fund (NWF): Background Report to the Blue Ribbon Commission on America's Nuclear Future." Blue Ribbon Commission.

Hiruo, Elaine. 2014. "DOE Implements Court-Ordered Suspension of Nuclear Waste Fee." *NuclearFuel*, May 26.

Holt, Mark. 2021. "Civilian Nuclear Waste Disposal." RL33461. Congressional Research Service.

Holtorf, Cornelius, and Anders Högberg. 2015. "Archaeology and the Future: Managing Nuclear Waste as a Living Heritage." In *Radioactive Waste Management and Constructing Memory for Future Generations. Proceedings of the International Conference and Debate, 15-17 September 2014, Verdun, France*. 97–101.

Hora, Stephen C., and Detlof Von Winterfeldt. 1997. "Nuclear Waste and Future Societies: A Look into the Deep Future." *Technological Forecasting and Social Change* 56 (2): 155–170.

Houdart, Sophie. 2017. "Les Répertoires Subtils d'un Terrain Contaminé." *Techniques & Culture. Revue Semestrielle d'anthropologie Des Techniques* 68: 88–103.

Houdart, Sophie. 2018. "Prendre Soi de Ce Qui Blesse. L'insoutenable 'vivre Avec' La Radioactivité Dans La Région de Fukushima." Penser le soin à travers ses Tensions: Fragilité et Force du Vivant et de Son Milieu, International Colloque organized by Laurence Tessier. Centre d'Etude des Mouvements Sociaux, Institut Marcel Mauss, EHESS, Paris.

Houston, Donna. 2013. "Environmental Justice Storytelling: Angels and Isotopes at Yucca Mountain, Nevada." *Antipode* 45 (2): 417–435.

Howard, John H., and Lawson E. Benfield. 1991. "Rulemaking in the Shadows: The Rise of OMB and Cost-Benefit Analysis in Environmental Decisionmaking." *Columbia Journal of Environmental Law* 16: 143–192.

Hyndman, Noel, and Ciaran Connolly. 2011. "Accruals Accounting in the Public Sector: A Road Not Always Taken." *Management Accounting Research* 22 (1): 36–45.

Ialenti, Vincent. 2020. *Deep Time Reckoning: How Future Thinking Can Help Earth Now.* MIT Press.

Ialenti, Vincent. 2021. "Drum Breach: Operational Temporalities, Error Politics and WIPP's Kitty Litter Nuclear Waste Accident." *Social Studies of Science* 51 (3): 364–391.

Interagency Review Group. 1978. "Report to the President by the Interagency Review Group on Nuclear Waste Management." TID-28817-Draft.

International Atomic Energy Agency. 1997. *Joint Convention on the Safety of Spent Fuel Management and on the Safety of Radioactive Waste Management.* International Atomic Energy Agency.

Irvine, Richard. 2014. "Deep Time: An Anthropological Problem." *Social Anthropology/Anthropologie Sociale* 22 (2): 157–172.

Irvine, Richard. 2020. *An Anthropology of Deep Time: Geological Temporality and Social Life.* Cambridge University Press.

Ishiyama, Noriko. 2003. "Environmental Justice and American Indian Tribal Sovereignty: Case Study of a Land–Use Conflict in Skull Valley, Utah." *Antipode* 35 (1): 119–139.

Ishiyama, Noriko, and Kim TallBear. 2022. "9 Nuclear Waste and Relational Accountability in Indian Country." In *The Promise of Multispecies Justice*, edited by Sophie Chao, Karin Bolender, and Eben Kirksey, 185–204. Duke University Press.

Jacob, Gerald. 1990. *Site Unseen: The Politics of Siting a Nuclear Waste Repository.* Pitt Series in Policy and Institutional Studies. University of Pittsburgh Press.

Janberg, Klaus, and Frank von Hippel. 2009. "Dry-Cask Storage: How Germany Led the Way." *Bulletin of the Atomic Scientists* 65 (5): 24–32.

Jasanoff, Sheila, and Sang-Hyun Kim. 2009. "Containing the Atom: Sociotechnical Imaginaries and Nuclear Power in the United States and South Korea." *Minerva* 47 (2): 119–146.

Jasanoff, Sheila, and Sang-Hyun Kim. 2015. *Dreamscapes of Modernity: Sociotechnical Imaginaries and the Fabrication of Power*. University of Chicago Press.

Johansson, Stina. 2020. "Tax." In *Cambridge Encyclopedia of Anthropology*, 1–16. Cambridge University Press.

Joyce, Rosemary. 2020. *The Future of Nuclear Waste: What Art and Archaeology Can Tell Us About Securing the World's Most Hazardous Material*. Oxford University Press.

Joyce, Rosemary. 2024. "Toxic and Wasted: Artists Thinking About How to Engage with Material Futures." In *Toxic Heritage* edited by Elisabeth Kryder-Reid and Sarah May, 332–342. Routledge.

Kalshoven, Petra T. 2022. "The Fish Is in the Water and the Water Is in the Fish: Symbiosis in A Nuclear Whale Fall." *Cultural Anthropology* 37 (2): 349–378.

Kaplan, Maureen F., and Mel Adams. 1986. "Using the Past to Protect the Future: Marking Nuclear Waste Disposal Sites." *Archaeology* 39 (5): 51–54.

Kasperson, Roger E. 1984. "Equity Issues in Radioactive Waste Management." U.S. Department of Energy Office of Scientific and Technical Information.

Keck, Frédéric. 2017. "Stockpiling as a Technique of Preparedness: Conserving the Past for an Unpredictable Future." In *Cryopolitics: Frozen Life in a Melting World*, 117–142. MIT Press.

Kenley, C. R., and Donald C. Armstead. 2004. "Discounting Models for Long-Term Decision Making." *Systems Engineering* 7 (1): 13–24.

Kenner, Alison, Aftab Mirzaei, and Christy Spackman. 2019. "Breathing in the Anthropocene: Thinking through Scale with Containment Technologies." *Cultural Studies Review* 25 (2): 153–171.

Klein, Christine A. 1996. "Treaties of Conquest: Property Rights, Indian Treaties, and the Treaty of Guadalupe Hidalgo." *New Mexico Law Review* 26: 201.

Kraft, Michael E., and Bruce B. Clary. 1991. "Citizen Participation and the NIMBY Syndrome: Public Response to Radioactive Waste Disposal." *Western Political Quarterly* 44 (2): 299–328.

Kuchinskaya, Olga. 2014. *The Politics of Invisibility: Public Knowledge about Radiation Health Effects after Chernobyl*. MIT Press.

Kuletz, Valerie. 1998. *The Tainted Desert: Environmental Ruin in the American West*. Routledge.

Kunreuther, Howard. 1995. *Voluntary Procedures for Siting Noxious Facilities: Lotteries, Auctions and Benefit-Sharing*. Wharton Risk Management and Decision Processes Center.

Kunreuther, Howard, Paul Kleindorfer, Peter J. Knez, and Rudy Yaksick. 1987. "A Compensation Mechanism for Siting Noxious Facilities: Theory and Experimental Design." *Journal of Environmental Economics and Management* 14 (4): 371–383.

Laporte, T., G. I. Rochlin, Daniel Metlay, and P. Windham. 1976. "Space Disposal of Nuclear Wastes. *Volume 1*: Socio-Political Aspects." University of California.

Las Vegas Sun. 2006. "Editorial: Domenici as Jekyll & Hyde." *Las Vegas Sun*, September 29.

Latour, Bruno. 1984. "The Powers of Association." *The Sociological Review* 32 (S1): 264–280.

Latour, Bruno. 1992. "Where Are the Missing Masses? The Sociology of a Few Mundane Artifacts." In *Shaping Technology/Building Society: Studies in Sociotechnical Change*, edited by Wiebe E. Bijker and John Law, 225–258. MIT Press.

Latour, Bruno. 1997. "Trains of Thought: Piaget, Formalism, and the Fifth Dimension." *Common Knowledge* 6: 170–170.

Latour, Bruno. 2005. *Reassembling the Social: An Introduction to Actor-Network-Theory*. Oxford University Press.

Latour, Bruno. 2017. "Anthropology at the Time of the Anthropocene: A Personal View of What Is to Be Studied." In *The Anthropology of Sustainability*, edited by Marc Brightman and Jerome Lewis, 35–49. Palgrave Macmillan.

Lehtonen, Markku. 2010. "Deliberative Decision-Making on Radioactive Waste Management in Finland, France and the UK: Influence of Mixed Forms of Deliberation in the Macro Discursive Context." *Journal of Integrative Environmental Sciences* 7 (3): 175–196.

Leivestad, Hege H., and Johanna Markkula. 2021. "Inside Container Economies." *Focaal* 2021 (89): 1–11.

Lezaun, Javier, Fabian Muniesa, and Signe Vikkelsø. 2013. "Provocative Containment and the Drift of Social-Scientific Realism." *Journal of Cultural Economy* 6 (3): 278–293.

Liboiron, Max. 2021. *Pollution Is Colonialism*. Duke University Press.

Linhardt, Dominique, and Fabian Muniesa. 2011. "Tenir lieu de politique: Le paradoxe des «politiques d'économisation»." *Politix* 95 (3): 7.

Little, Ben, and Alison Winch. 2017. "Generation: The Politics of Patriarchy and Social Change." *Soundings* 66 (66): 129–144.

Long, Jane C. S., and Rodney C. Ewing. 2004. "YUCCA MOUNTAIN: Earth-Science Issues at a Geologic Repository for High-Level Nuclear Waste." *Annual Review of Earth and Planetary Sciences* 32 (1): 363–401.

Lucero, Brian L. 2007. "Old Towns Challenged by the Boom Town: The Villages of the Middle Rio Grande Valley and the Albuquerque Tricentennial." *New Mexico Historical Review* 82 (1): 3.

Macfarlane, Allison. 2003. "Underlying Yucca Mountain: The Interplay of Geology and Policy in Nuclear Waste Disposal." *Social Studies of Science* 33 (5): 783–807.

Macfarlane, Allison, and Rodney C. Ewing, eds. 2006. *Uncertainty Underground: Yucca Mountain and the Nation's High-Level Nuclear Waste.* MIT Press.

MacKenzie, Michael K. 2013. "Future Publics: Long-Term Thinking and Farsighted Action in Democratic Systems." PhD Thesis, University of British Columbia.

MacKenzie, Michael K. 2021. *Future Publics: Democracy, Deliberation, and Future-Regarding Collective Action*. Oxford University Press.

MacKenzie, Michael K., Maija Setälä, and Simo Kyllönen, eds. 2023. *Democracy and the Future: Future-Regarding Governance in Democratic Systems*. Edinburgh University Press.

Madden, Gregory J., and Warren K. Bickel. 2010. *Impulsivity: The Behavioral and Neurological Science of Discounting*. American Psychological Association.

Makovicky, Nicolette, and Robert Smith, eds. 2023. *Beyond the Social Contract: An Anthropology of Tax*. Vol. 15. Berghahn.

Malinowski, Bronislaw. 1990. "Time-Reckoning in the Trobriands." In *The Sociology of Time*, edited by John Hassard, 203–18. Palgrave Macmillan.

Malone, Michael P., and Richard W. Etulain. 1989. *The American West: A Twentieth-Century History*. University of Nebraska Press.

Manaugh, Geoff. 2009. "Fossil Reactors." *BLDGBLOG* (blog). October 26. https://bldgblog.com/2009/10/fossil-reactors/

Manne, Alan S. 1995. "The Rate of Time Preference: Implications for the Greenhouse Debate." *Energy Policy* 23 (4–5): 391–394.

Mannheim, Karl. 1952. *Essays on the Sociology of Knowledge*. Routledge & Kegan Paul.

Marks, Stephen G. 2011. "Valuing the Future: Intergenerational Discounting, Its Problems, and a Modest Proposal." *Envtl. L. Rep. News & Analysis* 41: 10615.

Marshall, T. 1998a. *A Socioeconomic Impact Study of the Waste Isolation Pilot Plant Carlsbad, New Mexico*. Riverside Research and Associates.

Marshall, T. 1998b. *Carlsbad*. Riverside Research.

Martin, Isaac W., Ajay K. Mehrotra, and Monica Prasad. 2009. *The New Fiscal Sociology: Taxation in Comparative and Historical Perspective.* Cambridge University Press.

Mascaro, Lisa. 2009. "Reid: White House to Cut Off Yucca Funding." *Las Vegas Sun,* July 30.

Masco, Joseph. 2005. "A Notebook on Desert Modernism: From the Nevada Test Site to Liberace's Two-Hundred-Pound Suit." In *Histories of the Future,* edited by Susan Friend Harding and Daniel Rosenberg, 19–49. Duke University Press.

Masco, Joseph. 2006. *The Nuclear Borderlands: The Manhattan Project in Post-Cold War New Mexico.* Princeton University Press.

Maurer, Bill. 2008. "Moral Economies, Economic Moralities." In *Economics and Morality,* edited by Katherine E. Browne and B. Lynne Milgram, 257–270. Rowman & Littlefield.

Maurer, Bill. 2014. "Postscript: Is There Money in Credit?" *Consumption Markets & Culture* 17 (5): 512–518.

Mazmanian, Daniel, and David Morell. 1990. "The 'NIMBY" Syndrome: Facility Siting and the Failure of Democratic Discourse." In *Environmental Policy in the 1990s: Toward a New Agenda,* edited by Norman J. Vig and Michael E. Kraft, 233–250. CQ Press.

McCain, John. 2008. "Remarks by John McCain on His Comprehensive Plan for Energy Security." *John McCain 2008—John McCain for President* (blog).

McClure, Samuel M., Keith M. Ericson, David I. Laibson, George Loewenstein, and Jonathan D. Cohen. 2007. "Time Discounting for Primary Rewards." *Journal of Neuroscience* 27 (21): 5796–5804.

McCutcheon, Chuck. 2002. *Nuclear Reactions: The Politics of Opening a Radioactive Waste Disposal Site.* University of New Mexico Press.

Mead, Margaret. 1970. *Culture and Commitment: A Study of the Generation Gap Culture and Commitment.* Natural History Press.

Méadel, Cécile, and Vololona Rabeharisoa. 1999. *Consommateurs et Produits Alimentaires: La Construction Des Ajustements. Partie II: Le Consommateur Mis En Bouteille: L'équipée de Pampryl et Banga.* CSI.

Metlay, Daniel. 2000. "From Tin Roof to Torn Wet Blanket: Predicting and Observing Groundwater Movement at a Proposed Nuclear Waste Site." In *Prediction: Science, Decision Making, and the Future of Nature,* edited by Daniel Sarewitz, Roger A. Pielke, and Radford Byerly, 199–228. Island Press.

Meyer, Lukas H. 2004. *Justice in Time: Responding to Historical Injustice. Vol. 30.* Nomos Verlagsgesellschaft.

Millar, Kathleen M. 2018. *Reclaiming the Discarded: Life and Labor on Rio's Garbage Dump*. Duke University Press.

Miller, Peter. 1991. "Accounting Innovation beyond the Enterprise: Problematizing Investment Decisions and Programming Economic Growth in the UK in the 1960s." *Accounting, Organizations and Society* 16 (8): 733–762.

Miller, Peter. 1998. "The Margins of Accounting." *European Accounting Review* 7 (4): 605–621.

Miller, Peter, and Ted O'Leary. 2007. "Mediating Instruments and Making Markets: Capital Budgeting, Science and the Economy." *Accounting, Organizations and Society* 32 (7–8): 701–734.

Mitchell, Robert C., and Richard T. Carson. 1986. "Property Rights, Protest, and the Siting of Hazardous Waste Facilities." *The American Economic Review* 76 (2): 285–290.

Moisey, Andrew. 2012. "Considering the Desire to Mark Our Buried Nuclear Waste." *Qui Parle* 20 (2): 101–125.

Moisey, Andrew. 2017. "Permanent Negative Value: The Waste Isolation Pilot Plant." *Critical Inquiry* 43 (4): 861–892.

Moore, Michael J., and W. K. Viscusi. 1990. "Discounting Environmental Health Risks: New Evidence and Policy Implications." *Journal of Environmental Economics and Management* 18 (2): S51–S62.

Mora, Carl J. 1999. *Sandia and the Waste Isolation Pilot Plant 1974-1999*. Sandia National Laboratories.

Morrison, Edward R. 1998. "Judicial Review of Discount Rates Used in Regulatory Cost-Benefit Analysis." *University of Chicago Law Review* 65: 1333–1369.

Mugler, Johannes, Magnus Sebastian Johansson, and Robert Smith. 2024. *Anthropology and Tax: Ethnographies of Fiscal Relations*. Cambridge University Press.

Muniesa, Fabian. 2000. "Un Robot Walrasien. Cotation Électronique et Justesse de La Découverte Des Prix." *Politix* 13 (52): 121–154.

Muniesa, Fabian. 2011. "A Flank Movement in the Understanding of Valuation." *The Sociological Review* 59 (s2): 24–38.

Muniesa, Fabian. 2014. *The Provoked Economy: Economic Reality and the Performative Turn*. Routledge.

Muniesa, Fabian, Liliana Doganova, Horacio Ortiz, et al. 2017. *Capitalization: A Cultural Guide*. Presse des Mines.

Munn, Nancy D. 1992. "The Cultural Anthropology of Time: A Critical Essay." *Annual Review of Anthropology* 21: 93–123.

Nagourney, Adam. 2009. "G.O.P. Eyes Tough Task: Winning Reid's Seat." *New York Times*, May 25.

Nash, Gerald D. 1990. *The American West Transformed: The Impact of the Second World War*. University of Nebraska Press.

Nash, Gerald D. 1999. *The Federal Landscape: An Economic History of the Twentieth-Century West*. University of Arizona Press.

National Research Council. 1957. *The Disposal of Radioactive Waste on Land*. The National Academies Press.

Nevada Commission on Nuclear Projects. 2006. "Report and Recommendations of the Nevada Commission on Nuclear Projects." Presented to the Governor and Legislature of the State of Nevada, December 2006.

Newell, Richard G., and William A. Pizer. 2003. "Discounting the Distant Future: How Much Do Uncertain Rates Increase Valuations?" *Journal of Environmental Economics and Management* 46 (1): 52–71.

Newell, Sasha. 2018. "Uncontained Accumulation: Hidden Heterotopias of Storage and Spillage." *History and Anthropology* 29 (1): 37–41.

Nixon, Rob. 2011. *Slow Violence and the Environmentalism of the Poor*. Harvard University Press.

Nuclear Waste Policy Act of 1982. 1982. *42 U.S.C.*

Obama, Barack. 2007. "Barack Obama Explains Yucca Mountain Stance." *Las Vegas Sun*, May 20.

Office of the Inspector General. 2010. "Need for Enhanced Surveillance During the Yucca Mountain Shut Down." OAS-SR-10-01. US Department of Energy.

Office of the Inspector General, US Nuclear Regulatory Commission. 2011. "NRC Chairman's Unilateral Decision to Terminate NRC's Review of DOE Yucca Mountain Repository License Application." OIG Case No. 11–05. NRC.

Office of Management and Budget. 1972. "Discount Rates to Be Used in Evaluating Time-Distributed Costs and Benefits." Circular A-94. Office of Management and Budget.

Office of Nuclear Energy. 2022. "5 Fast Facts about Spent Nuclear Fuel." March 10, 2022. https://www.energy.gov/ne/articles/5-fast-facts-about-spent-nuclear-fuel

Office of Technology Assessment. 1982. "Managing Commercial High-Level Radioactive Waste." US Government Publishing Office.

O'Hara, Patrick. 1992. "The NIMBY Syndrome Meets the Preemption Doctrine: Federal Preemption of State and Local Restrictions on the Siting of Hazardous Waste Disposal Facilities." *Louisiana Law Review* 53: 229.

O'Hare, Michael. 1977. "'Not On My Block You Don't'-Facilities Siting and the Strategic Importance of Compensation." Massachusetts Institute of Technology Laboratory of Architecture and Planning.

O'Hare, Michael, and Debra Sanderson. 1993. "Facility Siting and Compensation: Lessons from the Massachusetts Experience." *Journal of Policy Analysis and Management* 12 (2): 364–376.

O'Hare, Patrick. 2019. "'The Landfill Has Always Borne Fruit': Precarity, Formalisation and Dispossession among Uruguay's Waste Pickers." *Dialectical Anthropology* 43 (1): 31–44.

Oreskes, Naomi. 2000. "Why Predict? Historical Perspectives on Prediction in Earth Science." In *Prediction: Science, Decision Making, and the Future of Nature*, edited by Daniel Sarewitz, Roger A. Pielke, and Radford Byerly, 23–40. Island Press.

Oreskes, Naomi, Kristin Shrader-Frechette, and Kenneth Belitz. 1994. "Verification, Validation, and Confirmation of Numerical Models in the Earth Sciences." *Science* 263 (5147): 641–646.

Pacific Northwest Laboratory. 1978. "Sensitivity of the Federal Fee for Managing Spent Fuel to Financial and Logistical Variations." Prepared for the US Department of Energy Under Contract No. EY-76-C-06-1830 PNL-2637.

Pacific Northwest Laboratory. 1979. "Comparison of Spent Fuel Management Fee Collection Alternatives." Prepared for the US Department of Energy under contract EY-76-C-06-1830 PNL-2826.

Parker, Christopher M. 2023. "Carceral Citizenship in Puerto Rico." *European Review of Latin American and Caribbean Studies* 116: 87–104.

Parker, Christopher M. 2025. *Carceral Citizens: Labor and Confinement in Puerto Rico.* University of Chicago Press.

Patashnik, Eric M. 2000. *Putting Trust in the US Budget: Federal Trust Funds and the Politics of Commitment.* Cambridge University Press.

Patinaux, Lény. 2017. "Enfouir Des Déchets Nucléaires Dans Un Monde Conflictuel. Une Histoire de La Démonstration de Sûreté de Projets de Stockage Géologique, En France (1982–2013)." PhD Thesis, EHESS.

Paxson, Heather. 2010. "Locating Value in Artisan Cheese: Reverse Engineering Terroir for New-World Landscapes." *American Anthropologist* 112 (3): 444–457.

Pearce, David W., Giles Atkinson, and Susana Mourato. 2006. *Analyse Coûts-Bénéfices et Environnement: Développements Récents*. Organisation de coopération et développement économiques.

Pels, Peter. 2015. "Modern Times: Seven Steps toward an Anthropology of the Future." *Current Anthropology* 56 (6): 779–796. https://doi.org/10.1086/684047.

Petryna, Adriana. 2013. *Life Exposed.* Princeton University Press.

Pilmis, Olivier. Forthcoming. "Deep Focusing." *American Journal of Sociology.*

Pitkanen, Laura, and Matthew Farish. 2017. "Nuclear Landscapes." *Progress in Human Geography* 42 (6): 862–880.

Polivy, Janet, and C. P. Herman. 1992. "Undieting: A Program to Help People Stop Dieting." *International Journal of Eating Disorders* 11 (3): 261–268.

Pollitt, Christopher, and Geert Bouckaert. 2011. *Continuity and Change in Public Policy and Management.* Edward Elgar Publishing.

Porter, Theodore M. 1995. *Trust in Numbers: The Pursuit of Objectivity in Science and Public Life.* Princeton University Press.

Portney, Kent E. 1985. "The Potential of the Theory of Compensation for Mitigating Public Opposition to Hazardous Waste Treatment Facility Siting: Some Evidence from Five Massachusetts Communities." *Policy Studies Journal* 14 (1): 81.

Public Citizens. 2002. "Yucca Mountain Bought and Sold. Science Smothered under a Mountain of Nuclear Lobbyists." Public Citizens.

Quattrone, Paolo. 2004. "Accounting for God: Accounting and Accountability Practices in the Society of Jesus (Italy, XVI–XVII Centuries)." *Accounting, Organizations and Society* 29 (7): 647–683.

Quattrone, Paolo. 2005. "Is Time Spent, Passed or Counted? The Missing Link between Time and Accounting History." *Accounting Historians Journal* 32 (1): 185–208.

Radin, Joanna, and Emma Kowal, eds. 2017. *Cryopolitics: Frozen Life in a Melting World.* MIT Press.

Rae, John. 1905. *The Sociological Theory of Capital: Being a Complete Reprint of the New Principles of Political Economy, 1834.* Macmillan.

Ralston, John. 2008. "McCain's about-Face About Yucca." *Las Vegas Sun*, May 28.

Randle, Stephanie. 2024. "Geographies of Storage." *Geography Compass* 18 (1): e12733.

Reno, Joshua. 2009. "Your Trash Is Someone's Treasure: The Politics of Value at a Michigan Landfill." *Journal of Material Culture* 14 (1): 29–46.

Reno, Joshua. 2016. *Waste Away: Working and Living with a North American Landfill.* University of California Press.

Reno, Joshua O. 2020. *Military Waste: The Unexpected Consequences of Permanent War Readiness.* University of California Press.

Richter, Jennifer. 2017. "Energopolitics and Nuclear Waste: Containing the Threat of Radioactivity." *Energy Research & Social Science* 30, 61–70.

Richter, Jennifer, Michael J. Bernstein, and Mahmud Farooque. 2022. "The Process to Find a Process for Governance: Nuclear Waste Management and Consent-Based Siting in the United States." *Energy Research & Social Science* 87: 102473.

Rick, Scott, and George Loewenstein. 2008. "Intangibility in Intertemporal Choice." *Philosophical Transactions of the Royal Society B: Biological Sciences* 363 (1511): 3813–3824.

Robinson, Kim Stanley. 2020. *The Ministry for the Future: A Novel*. Orbit.

Rogers, Keith. 2009. "Anti-Yucca Mountain Strategist Catching Up." *Las Vegas Review-Journal*, February 2.

Rogers, Kenneth, and Marvin Kingsley. 2004. "The Politics of Interim Radioactive Waste Storage: The United States." *Environmental Politics* 13 (3): 590–611.

Roitman, Janet. 2014. *Anti-Crisis*. Duke University Press.

Rose, Michael. 2024. "Can We Represent Future Generations in Myopic Democracies?—Analyzing the Design, Feasibility and Viability of Institutional Guardians of Future Generations." *Policy Studies, December*, 1–21.

Rose, Nikolas. 1996. "Identity, Genealogy, History." In *Questions of Cultural Identity*, edited by S. Hall and P. du Gay, 128–150. Sage.

Rose, Nikolas, and Peter Miller. 1992. "Political Power beyond the State: Problematics of Government." *The British Journal of Sociology* 43 (2): 173–205.

Rosenberg, Daniel, and Susan Harding, eds. 2005. *Histories of the Future*. Duke University Press.

Roth, Emile. 1977. "The Discovery and Study of the Nuclear Reactor in Oklo." *Journal of Radioanalytical Chemistry* 37 (1): 65–78.

Rowell, Arden. 2010. "The Cost of Time: Haphazard Discounting and the Undervaluation of Regulatory Benefits." *Notre Dame Law Review* 85 (4): 10–30.

Sachs, Noah. 1996. "The Mescalero Apache and Monitored Retrievable Storage of Spent Nuclear Fuel: A Study in Environmental Ethics." *Natural Resources Journal* 36: 641.

Saraç-Lesavre, Başak. 2015. "Formulating Nuclear Values: Communities, Equations, Budgets and Debates with Nuclear Waste." PhD diss., Ecole des Mines de Paris, Centre de Sociologie de l'Innovation.

Saraç-Lesavre, Başak. 2018. "Budgeting the Future: Negotiating the Values of a Contemporary Liberal Democracy." In *Materiality and Managerial Techniques*, edited by Nathalie Mitev, Anna Morgan-Thomas, Philippe Lorino, Francois-Xavier De Vaujany, and Yesh Nama, 57–76. Springer International Publishing.

Saraç-Lesavre, Başak. 2021a. "Deep Time Financing? 'Generational' responsibilities and the Problem of Rendez-Vous in the US Nuclear Waste Programme." *Journal of Cultural Economy* 14 (4): 435–448.

Saraç-Lesavre, Başak. 2021b. "Deep Time: The End of an Engagement." *Issues in Science and Technology* 37 (3): 70–72.

Saraç-Lesavre, Başak. 2021c. "Styles of Revaluation: The Case of the Levelized Cost of Electricity." *Nuclear Technology* 207 (9): 1366–1376.

Saraç-Lesavre, Basak. Forthcoming. "Navigating the Nuclear Divide: Defense and Civilian High-Level Waste Management."

Sarewitz, Daniel R., Roger A. Pielke, and Radford Byerly. 2000. *Prediction: Science, Decision Making, and the Future of Nature*. Island Press.

Scheffler, Samuel. 2018. *Why Worry about Future Generations?* Oxford University Press.

Schelling, Thomas C. 1995. "Intergenerational Discounting." *Energy Policy* 23 (4): 395–401.

Schick, Allen. 2000. *The Federal Budget: Politics, Policy, Process*. Brookings Institution Press.

Schoot, Ignace, and Charles Mather. 2022. "Opening Up Containment." *Science, Technology, & Human Values* 47 (5): 937–959.

Schwartz, Stephen I. 2011. *Atomic Audit: The Costs and Consequences of US Nuclear Weapons Since 1940*. Brookings Institution Press.

Sébastien, Léa. 2013. "Le nimby est mort. Vive la résistance éclairée: le cas de l'opposition à un projet de décharge, Essonne, France." *Sociologies Pratiques* 27 (2): 145–165.

Sebeok, T. A. 1984. *Communication Measures to Bridge Ten Millennia*. Office of Nuclear Waste Isolation, Battelle Memorial Institute.

Shrader-Frechette, Kristin S. 1993. "Risk Society: Towards a New Modernity." *Economic Geography* 69 (4): 432.

Shrader-Frechette, Kristin. 2005. "Mortgaging the Future: Dumping Ethics with Nuclear Waste." *Science and Engineering Ethics* 11 (4): 518–520.

Simmel, Georg. 1950. *The Sociology of Georg Simmel: Trans*. Free Press.

Simmons, Peter, and Karen Bickerstaff. 2006. "The Participatory Turn in UK Radioactive Waste Management Policy." In Proceedings of VALDOR-2006. Congrex Sweden, AB, 529–536.

Simonich, Milan. 2013. "Carlsbad Attorney Roxanne Lara Says She's Running for Congress." *Carlsbad Current-Argus*, March 9.

Simpson, Michael. 2019. "The Annihilation of Time by Space: Pluri-temporal Strategies of Capitalist Circulation." *Environment and Planning E: Nature and Space* 2 (1): 110–128.

Siné, Alexandre. 2006. *L'ordre Budgétaire. L'économie Politique Des Dépenses de l'État*. Economica.

Smith, Graham. 2020. "Enhancing the Legitimacy of Offices for Future Generations: The Case for Public Participation." *Political Studies* 68 (4): 996–1013.

Smith, Jessica, and Mette M. High. 2017. "Exploring the Anthropology of Energy: Ethnography, Energy and Ethics." *Energy Research & Social Science* 30: 1–6.

Sofia, Zoe. 2000. "Container Technologies." *Hypatia* 15 (2): 181–201.

Spackman, Michael. 2004. "Time Discounting and of the Cost of Capital in Government." *Fiscal Studies* 25 (4): 467–518.

Spears, Taylor. 2019. "Discounting Collateral: Quants, Derivatives and the Reconstruction of the 'Risk-Free Rate' after the Financial Crisis." *Economy and Society* 48 (3): 342–370.

Spence, Crawford. 2010. "Accounting for the Dissolution of a Nation State: Scotland and the Treaty of Union." *Accounting, Organizations and Society* 35 (3): 377–392.

Stahl, David. 2006. "Dripshields and Backfill." In *Uncertainty Underground: Yucca Mountain and the Nation's High-Level Nuclear Waste*, edited by Allison Macfarlane and Rodney C. Ewing. MIT Press.

Star, Susan L. 1999. "The Ethnography of Infrastructure." *American Behavioral Scientist* 43 (3): 377–391.

Stark, David, and Michael Hutter. 2015. "Pragmatist Perspectives on Valuation: An Introduction." In *Moments of Valuation: Exploring Sites of Dissonance*, edited by Ariane Berthoin Antal, Michael Hutter, and David Stark, 1–12. Oxford University Press.

Stawkowski, Magdalena E. 2016. "'I Am a Radioactive Mutant': Emergent Biological Subjectivities at Kazakhstan's Semipalatinsk Nuclear Test Site." *American Ethnologist* 43 (1): 144–157.

Stewart, Richard B., and Jane B. Stewart. 2011. *Fuel Cycle to Nowhere: U.S. Law and Policy on Nuclear Waste*. Vanderbilt University Press.

Stoler, Ann L. 2016. *Duress: Imperial Durabilities in Our Times*. Duke University Press.

Sundqvist, Göran, and Mark Elam. 2010. "Public Involvement Designed to Circumvent Public Concern? The 'Participatory Turn' in European Nuclear Activities." *Risk, Hazards & Crisis in Public Policy* 1 (4): 203–229.

Sunstein, Cass R., and Arden Rowell. 2007. "On Discounting Regulatory Benefits: Risk, Money, and Intergenerational Equity." *The University of Chicago Law Review* 74: 171–208.

Szasz, Ferenc. 1998. "New Mexico's Forgotten Nuclear Tests: Projects Gnome (1961) and Gasbuggy (1967)." *New Mexico Historical Review* 73, 4.

Tănăsescu, Mihnea. 2020. "Rights of Nature, Legal Personality, and Indigenous Philosophies." *Transnational Environmental Law* 9 (3): 429–453.

Teuscher, Ursina, and Suzanne H. Mitchell. 2011. "Relation between Time Perspective and Delay Discounting: A Literature Review." *The Psychological Record* 61 (4): 613–632.

The Associated Press. 2006. "Domenici Offers Nuclear Waste Proposal." *USA Today*, June 27, 2006.

The Carlsbad Mayor's Nuclear Task Force. 2012. "The Nuclear Nexus." March.

The White House. n.d. "The OMB Guidelines Section 20. Terms and Concepts." https://www.whitehouse.gov/omb/information-for-agencies/circulars

Thompson, Dennis F. 2015. "Representing Future Generations: Political Presentism and Democratic Trusteeship." In *Democracy, Equality, and Justice*, edited by Matt Matravers, 17–38. Routledge.

Thompson, Edward P. 1971. "The Moral Economy of the English Crowd in the Eighteenth Century." *Past & Present* 50 (1): 76–136.

Thompson, Janna. 2009. *Intergenerational Justice: Rights and Responsibilities in an Intergenerational Polity*. Routledge.

Trom, Danny. 1999. "De la réfutation de l'effet NIMBY considérée comme une pratique militante. Notes pour une approche pragmatique de l'activité revendicative." *Revue française de science politique* 49 (1): 31–50.

Trompette, Pascale. 2013. "The Politics of Value in French Funeral Arrangements: Three Types of Moral Calculation." *Journal of Cultural Economy* 6 (4): 370–385.

Trompette, Pascale, and Robert H. Griffiths. 2011. "L'économie Morale de La Mort Au XIXe Siècle. Regards Croisés Sur La France et l'Angleterre." *Le Mouvement Social* 237 (4): 33–54.

Trundle, Catherine. 2011. "Biopolitical Endpoints: Diagnosing a Deserving British Nuclear Test Veteran." *Social Science & Medicine* 73 (6): 882–888.

TRW Energy Systems Planning Division. 1978. "Report of Task Force for Review of Nuclear Waste Management." DOE/ER-0004/D-Draft. US Department of Energy.

Tsing, Anna L. 2015. *The Mushroom at the End of the World: On the Possibility of Life in Capitalist Ruins*. Princeton University Press.

Tsing, Anna L., Nils Bubandt, Elaine Gan, and Heather A. Swanson. 2017. *Arts of Living on a Damaged Planet: Ghosts and Monsters of the Anthropocene*. University of Minnesota Press.

Tuler, Seth P., and Thomas Webler. 2023. "Controversy and Consensus in the Design of a Consent-Based Siting Process for Radioactive Waste." *Energy Research & Social Science* 95: 102906.

Uribe, J., N. Saraeva, K. Petry, E. Bickford, R. Kreuzer, and R. Howard. 2024. "Current Progress on the Development of the Integrated Waste Management Consent-Based Siting Process in the United States." *Nuclear Technology* 210 (9): 1748–1753.

US Congressional Budget Office. 1981. "Financing Radioactive Waste Disposal." CBO Staff Working Paper.

US Congressional Budget Office. 1982. "Financing Radioactive Waste Disposal." Y 10.2:F 49/2. US Government Printing Office.

US Department of Energy. 1978. "Preliminary Estimates of the Charge for Spent Fuel Storage and Disposal Services." DOE/ET-0055. US Government Printing Office.

US Department of Energy. 1980. "Report on the Fee for Spent Nuclear Fuel Storage and Disposal Services." DOE/S R-0006/UC -85. US Government Printing Office.

US Department of Energy. 1983a. "Nuclear Waste Policy Act Project Office, Report on Financing the Disposal of Commercial Spent Nuclear Fuel and Processed High-Level Radioactive Waste." DOE/5-0020/1. US Congress.

US Department of Energy. 1983b. "Report on Financing the Disposal of Commercial Spent Nuclear Fuel and Processed High-Level Radioactive Waste." DOEIS-0020/1. US Department of Energy.

US Department of Energy. 1984. "Managing Nuclear Waste: A Better Idea." US Department of Energy.

US Department of Energy. 1985. "Response Letter to the Report of the Panel on the Alternative Means for Managing and Financing Civilian Radioactive Waste Policy," April 18, 1985. US Department of Energy.

US Department of Energy. 1998. "Nuclear Waste Fund Fee Adequacy: An Assessment." US DOE/RW-0509. US Department of Energy.

US Department of Energy. 1999. "A Roadmap for Developing Accelerator Transmutation of Waste (ATW) Technology. A Report to Congress." DOE/RW-0519. US Department of Energy.

US Department of Energy. 2001. "Alternative Means of Financing and Managing the Civilian Radioactive Waste Management Program." DOE/RW-0546. US Department of Energy.

US Department of Energy. 2008. "Fiscal Year 2007 Civilian Radioactive Waste Management Fee Adequacy Assessment Report." DOE/RW-0593. US Department of Energy.

US Department of Energy, Office of Environmental Management. 2015. "Accident Investigation Report. Phase 2. Radiological Release Event at the Waste Isolation Pilot Plant, February 14, 2014." U.S. Department of Energy, April 2015.

US Department of Energy, Office of Nuclear Energy. 2007. "Global Nuclear Energy Partnership. Programmatic Environmental Impact Statement. Public Hearing and

Invitation to Comment February 27, 2007. Carlsbad, New Mexico." Bean & Associates, Inc.

US Department of Energy, Office of Nuclear Energy. 2008a. "Global Nuclear Energy Partnership. Programmatic Environmental Impact Statement. Public Hearing and Invitation to Comment November 18, 2008. Carlsbad, New Mexico." Bean & Associates, Inc.

US Department of Energy, Office of Nuclear Energy. 2008b. "Global Nuclear Energy Partnership. Programmatic Environmental Impact Statement. Public Hearing and Invitation to Comment November 18, 2008. Hobbs, New Mexico." Bean & Associates, Inc.

US Department of Energy, Office of Nuclear Energy. 2008c. "Global Nuclear Energy Partnership. Programmatic Environmental Impact Statement. Public Hearing and Invitation to Comment November 18, 2008. Roswell, New Mexico." Bean & Associates, Inc.

US Department of Energy, Office of Nuclear Energy. 2008d. "Global Nuclear Energy Partnership. Programmatic Environmental Impact Statement. Public Hearing and Invitation to Comment November 20, 2008. Los Alamos, New Mexico." Bean & Associates, Inc.

US Department of Energy, Carlsbad Field Office. 2012. "A Conceptual Plan for Salt Defense Disposal Investigations for the Disposal of DOE-EM Managed Wastes." DOE/CBFO-12-3485. US Department of Energy.

US Government Accountability Office. 2011a. "Better Information Needed on Waste Storage at DOE Sites as a Result of Yucca Mountain Shut Down." GAO-11-230. US Government Printing Office.

US Government Accountability Office. 2011b. "Commercial Nuclear Waste: Effects of a Termination of the Yucca Mountain Repository Program and Lessons Learned." GAO-11-229. US Government Printing Office.

US Government Accountability Office. 2011c. "Yucca Mountain: Information on Alternative Uses of the Site and Related Challenges." Report to the Majority Leader, US Senate; GAO-11-847. US Government Printing Office.

US Government Accountability Office. 2012. "Spent Nuclear Fuel: Accumulating Quantities at Commercial Reactors Present Storage and Other Challenges." GAO-12-797. US Government Printing Office.

US Government Accountability Office. 2022. "Financial Audit: FY 2021 and FY 2020 Consolidated Financial Statements of the US Government." GAO-22-105122. GAO.

US Government Accountability Office. 2025. "Nuclear Waste Cleanup: DOE Needs to Improve Contractor Oversight at the Waste Isolation Pilot Plant." GAO-25-107333. US Government Accountability Office.

US Government Accounting Office. 1993. "Nuclear Waste Yucca Mountain Project Behind Schedule and Facing Major Scientific Uncertainties." GA 1.13:RCED-93-124. US Government Publishing Office.

US Government Accounting Office, Comptroller General. 1986. "Department of Energy's Initial Efforts to Implement the Nuclear Waste Policy Act of 1982."

US Nuclear Regulatory Commission. 1978. "Draft Generic Environmental Impact Statement on Handling and Storage of Spent Light Water Reactor Fuel." NUREG-0404. US Nuclear Regulatory Commission.

US Nuclear Regulatory Commission. 2010. "Congressional Budget Justification for FY 2011." Nureg-1100, *Volume 26.*

US Nuclear Regulatory Commission. 2014. *Continued Storage of Spent Nuclear Waste.* US Nuclear Regulatory Commission.

US Nuclear Regulatory Commission. 2025. "Definitions." Code of Federal Regulations, title 10, part 60, section 60.2. "Disposal of High-Level Radioactive Wastes in Geologic Repositories." US Nuclear Regulatory Commission.

US Nuclear Regulatory Commission, Chief Financial Officer. 2010. "Guidance Under A Fiscal Year 2011 Continuing Resolution." US Nuclear Regulatory Commission.

Van Wyck, Peter C. 1997. "Signs of Danger/dangerous Signs: Responding to Nuclear Threat." PhD diss., Department of Art History and Communication Studies, McGill University.

VandeHei, Jim. 2004. "Kerry Has Nevada's Ear on Yucca Mountain Plan." *Washington Post,* August 11, 2004.

Vatin, François. 2007. *Morale Industrielle et Calcul Économique Dans Le Premier XIXème Siècle: L'économie Industrielle de Claude-Lucien Bergery (1787–1863).* L'Harmattan.

Vatin, François. 2008. "L'esprit d'ingénieur: pensée calculatoire et éthique économique." *Revue Française de Socio-Économie* 1 (1): 131.

Vatin, François, ed. 2013. *Évaluer et valoriser une sociologie économique de la mesure.* Presses universitaires du Mirail.

Venkatesan, Soumhya. 2020. "Afterword: Putting Together the Anthropology of Tax and the Anthropology of Ethics." *Social Analysis* 64 (2): 141–154.

Viscusi, W. K. 2000. "Risk Equity." *The Journal of Legal Studies* 29 (S2): 843–871.

Viscusi, W. K. 2006. "Rational Discounting for Regulatory Analysis." Working paper no. 12294. National Bureau of Economic Research.

Viscusi, W. K. 2007. "Rational Discounting for Regulatory Analysis." *University of Chicago Law Review* 74: 209.

Viscusi, W. K., and Joseph E. Aldy. 2003. "The Value of a Statistical Life: A Critical Review of Market Estimates throughout the World." *Journal of Risk and Uncertainty* 27 (1): 5–76.

Viscusi, W. K., and Richard Zeckhauser. 1976. "Environmental Policy Choice under Uncertainty." *Journal of Environmental Economics and Management* 3 (2): 97–112.

Voyles, Traci B. 2015. *Wastelanding: Legacies of Uranium Mining in Navajo Country*. University of Minnesota Press.

Walker, J. Samuel. 2009. *The Road to Yucca Mountain: The Development of Radioactive Waste Policy in the United States*. University of California Press.

Wallace, Anthony FC. 1956. "Revitalization Movements." *American Anthropologist* 58 (2): 264–281.

Weitzman, Martin L. 1998. "Why the Far-Distant Future Should Be Discounted at Its Lowest Possible Rate." *Journal of Environmental Economics and Management* 36 (3): 201–208.

Weitzman, Martin L. 2001. "Gamma Discounting." *The American Economic Review* 9 (1): 260–271.

Weitzman, Martin L. 2009. "On Modeling and Interpreting the Economics of Catastrophic Climate Change." *The Review of Economics and Statistics* 91 (1): 1–19.

Weston, Kath. 2017. *Animate Planet: Making Visceral Sense of Living in a High-Tech Ecologically Damaged World*. Duke University Press.

Wexler, Mark N. 1996. "A Sociological Framing of the NIMBY (Not-in-My-Backyard) Syndrome." *International Review of Modern Sociology* 26 (1): 91–110.

Wheaton, Sara. 2008. "Obama Ad Attacks McCain on Yucca Mountain." *The Caucus, The Politics and Government Blog of the New York Times* (blog). August 9.

White, Jonathan. 2013. "Thinking Generations." *The British Journal of Sociology* 64 (2): 216–247.

White, Jonathan. 2017. "Climate Change and the Generational Timescape." *The Sociological Review* 65 (4): 763–778.

White, Jonathan. 2024. "Technocratic Myopia: On the Pitfalls of Depoliticising the Future." *European Journal of Social Theory* 27 (2): 260–278.

White, Joseph, and Aaron B. Wildavsky. 1989. *The Deficit and the Public Interest: The Search for Responsible Budgeting in the 1980s*. University of California Press.

Wildavsky, Aaron B. 1988. *The New Politics of the Budgetary Process*. Scott Foresman.

Wolsink, Maarten. 1994. "Entanglement of Interests and Motives: Assumptions behind the NIMBY-Theory on Facility Siting." *Urban Studies* 31 (6): 851–866.

Wolsink, Maarten. 2006. "Invalid Theory Impedes Our Understanding: A Critique on the Persistence of the Language of NIMBY." *Transactions of the Institute of British Geographers* 31 (1): 85–91.

Yandle, Bruce. 1991. "User Charges, Rent Seeking, and Public Choice." In *Charging for Government (Routledge Revivals): User Charges and Earmarked Taxes in Principle and Practice*, edited by Richard E. Wagner, 34. Routledge.

Yehle, E. 2010. "Yucca Project's Last 600 Employees Scramble for New Jobs." *New York Times*, August 4.

Yon, Guillaume, and Alexandre Mallard. 2012. "The Moral Life of Formulas. The Construction of Marginal Cost Pricing of Electricity in Postwar France." Presented at the conference: Embeddedness and Beyond: Do Sociological Theories Meet Economic Realities?, Moscow.

Youatt, Rafi. 2017. "Personhood and the Rights of Nature: The New Subjects of Contemporary Earth Politics." *International Political Sociology* 11 (1): 39–54.

Yusoff, Kathryn. 2018. *A Billion Black Anthropocenes or None*. University of Minnesota Press.

Zeckhauser, Richard. 2006. "Investing in the Unknown and Unknowable." *Capitalism and Society* 1 (2). http://www.bepress.com/cas/vol1/iss2/art5.

Zeckhauser, Richard J., and W. K. Viscusi. 2008. "Discounting Dilemmas: Editors' Introduction." *Journal of Risk and Uncertainty* 37 (2): 95–106.

Zelizer, Viviana A. R. 2017. *Morals and Markets: The Development of Life Insurance in the United States*. Columbia University Press.

Zerubavel, Eviatar. 1982. "The Standardization of Time: A Sociohistorical Perspective." *American Journal of Sociology* 88 (1): 1–23.

Zonabend, Françoise. 2007. *The Nuclear Peninsula*. Cambridge University Press.

Index

Inside Technology Series

Edited by Wiebe E. Bijker, Edward Jones-Imhotep, and Rebecca Slayton

Başak Saraç-Lesavre, *Nuclear Remains: On Temporalities, Responsibilities, and Values*

Peter F. Peters, *Electronic Baroque: Building a Historical Organ for the Present*

John Downer, *Rational Accidents: Reckoning with Catastrophic Technologies*

Jesse Olszynko-Gryn, *A Woman's Right to Know: Pregnancy Testing in Twentieth-Century Britain*

Dagmar Schäfer, Annapurna Mamidipudi, and Marius Buning, editors, *Ownership of Knowledge: Beyond Intellectual Property*

Tamar Novick, *Milk and Honey: Technologies of Plenty in the Making of a Holy Land*

Kristin Asdal and Tone Huse, *Nature-made Economy: Cod, Capital, and the Great Transformation of the Ocean*

Cyrus C. M. Mody, *The Squares: US Physical and Engineering Scientists in the Long 1970s*

Brice Laurent, *European Objects: The Troubled Dreams of Harmonization*

Florian Jaton, *The Constitution of Algorithms: Ground-Truthing, Programming, Formulating*

Kean Birch and Fabian Muniesa, *Turning Things into Assets*

David Demortain, *The Science of Bureaucracy: Risk Decision-Making and the US Environmental Protection Agency*

Nancy Campbell, *OD: Naloxone and the Politics of Overdose*

Lukas Engelmann and Christos Lynteris, *Sulphuric Utopias: The History of Maritime Fumigation*

Zara Mirmalek, *Making Time on Mars*

Joeri Bruyninckx, *Listening in the Field: Recording and the Science of Birdsong*

Edward Jones-Imhotep, *The Unreliable Nation: Hostile Nature and Technological Failure in the Cold War*

Jennifer L. Lieberman, *Power Lines: Electricity in American Life and Letters, 1882–1952*

Jess Bier, *Mapping Israel, Mapping Palestine: Occupied Landscapes of International Technoscience*

Benoît Godin, *Models of Innovation: The History of an Idea*

Stephen Hilgartner, *Reordering Life: Knowledge and Control in the Genomics Revolution*

Brice Laurent, *Democratic Experiments: Problematizing Nanotechnology and Democracy in Europe and the United States*

Cyrus C. M. Mody, *The Long Arm of Moore's Law: Microelectronics and American Science*

Tiago Saraiva, *Fascist Pigs: Technoscientific Organisms and the History of Fascism*

Teun Zuiderent-Jerak, *Situated Interventions: Sociological Experiments in Healthcare*

Basile Zimmermann, *Technology and Cultural Difference: Electronic Music Devices, Social Networking Sites, and Computer Encodings in Contemporary China*

Andrew J. Nelson, *The Sound of Innovation: Stanford and the Computer Music Revolution*

Sonja D. Schmid, *Producing Power: The Pre-Chernobyl History of the Soviet Nuclear Industry*

Casey O'Donnell, *Developer's Dilemma: The Secret World of Videogame Creators*

Christina Dunbar-Hester, *Low Power to the People: Pirates, Protest, and Politics in FM Radio Activism*

Eden Medina, Ivan da Costa Marques, and Christina Holmes, editors, *Beyond Imported Magic: Essays on Science, Technology, and Society in Latin America*

Anique Hommels, Jessica Mesman, and Wiebe E. Bijker, editors, *Vulnerability in Technological Cultures: New Directions in Research and Governance*

Amit Prasad, *Imperial Technoscience: Transnational Histories of MRI in the United States, Britain, and India*

Charis Thompson, *Good Science: The Ethical Choreography of Stem Cell Research*

Tarleton Gillespie, Pablo J. Boczkowski, and Kirsten A. Foot, editors, *Media Technologies: Essays on Communication, Materiality, and Society*

Catelijne Coopmans, Janet Vertesi, Michael Lynch, and Steve Woolgar, editors, *Representation in Scientific Practice Revisited*

Rebecca Slayton, *Arguments that Count: Physics, Computing, and Missile Defense, 1949–2012*

Stathis Arapostathis and Graeme Gooday, *Patently Contestable: Electrical Technologies and Inventor Identities on Trial in Britain*

Jens Lachmund, *Greening Berlin: The Co-Production of Science, Politics, and Urban Nature*

Chikako Takeshita, *The Global Biopolitics of the IUD: How Science Constructs Contraceptive Users and Women's Bodies*

Cyrus C. M. Mody, *Instrumental Community: Probe Microscopy and the Path to Nanotechnology*

Morana Alač, *Handling Digital Brains: A Laboratory Study of Multimodal Semiotic Interaction in the Age of Computers*

Gabrielle Hecht, editor, *Entangled Geographies: Empire and Technopolitics in the Global Cold War*

Michael E. Gorman, editor, *Trading Zones and Interactional Expertise: Creating New Kinds of Collaboration*

Matthias Gross, *Ignorance and Surprise: Science, Society, and Ecological Design*

Andrew Feenberg, *Between Reason and Experience: Essays in Technology and Modernity*

Wiebe E. Bijker, Roland Bal, and Ruud Hendricks, *The Paradox of Scientific Authority: The Role of Scientific Advice in Democracies*

Park Doing, *Velvet Revolution at the Synchrotron: Biology, Physics, and Change in Science*

Gabrielle Hecht, *The Radiance of France: Nuclear Power and National Identity after World War II*

Richard Rottenburg, *Far-Fetched Facts: A Parable of Development Aid*

Michel Callon, Pierre Lascoumes, and Yannick Barthe, *Acting in an Uncertain World: An Essay on Technical Democracy*

Ruth Oldenziel and Karin Zachmann, editors, *Cold War Kitchen: Americanization, Technology, and European Users*

Deborah G. Johnson and Jameson W. Wetmore, editors, *Technology and Society: Building Our Sociotechnical Future*

Trevor Pinch and Richard Swedberg, editors, *Living in a Material World: Economic Sociology Meets Science and Technology Studies*

Christopher R. Henke, *Cultivating Science, Harvesting Power: Science and Industrial Agriculture in California*

Helga Nowotny, *Insatiable Curiosity: Innovation in a Fragile Future*

Karin Bijsterveld, *Mechanical Sound: Technology, Culture, and Public Problems of Noise in the Twentieth Century*

Peter D. Norton, *Fighting Traffic: The Dawn of the Motor Age in the American City*

Joshua M. Greenberg, *From Betamax to Blockbuster: Video Stores and the Invention of Movies on Video*

Mikael Hård and Thomas J. Misa, editors, *Urban Machinery: Inside Modern European Cities*

Christine Hine, *Systematics as Cyberscience: Computers, Change, and Continuity in Science*

Wesley Shrum, Joel Genuth, and Ivan Chompalov, *Structures of Scientific Collaboration*

Shobita Parthasarathy, *Building Genetic Medicine: Breast Cancer, Technology, and the Comparative Politics of Health Care*

Kristen Haring, *Ham Radio's Technical Culture*

Atsushi Akera, *Calculating a Natural World: Scientists, Engineers and Computers during the Rise of U.S. Cold War Research*

Donald MacKenzie, *An Engine, Not a Camera: How Financial Models Shape Markets*

Geoffrey C. Bowker, *Memory Practices in the Sciences*

Christophe Lécuyer, *Making Silicon Valley: Innovation and the Growth of High Tech, 1930–1970*

Anique Hommels, *Unbuilding Cities: Obduracy in Urban Sociotechnical Change*

David Kaiser, editor, *Pedagogy and the Practice of Science: Historical and Contemporary Perspectives*

Charis Thompson, *Making Parents: The Ontological Choreography of Reproductive Technology*

Pablo J. Boczkowski, *Digitizing the News: Innovation in Online Newspapers*

Dominique Vinck, editor, *Everyday Engineering: An Ethnography of Design and Innovation*

Nelly Oudshoorn and Trevor Pinch, editors, *How Users Matter: The Co-Construction of Users and Technology*

Peter Keating and Alberto Cambrosio, *Biomedical Platforms: Realigning the Normal and the Pathological in Late-Twentieth-Century Medicine*

Paul Rosen, *Framing Production: Technology, Culture, and Change in the British Bicycle Industry*

Maggie Mort, *Building the Trident Network: A Study of the Enrollment of People, Knowledge, and Machines*

Donald MacKenzie, *Mechanizing Proof: Computing, Risk, and Trust*

Geoffrey C. Bowker and Susan Leigh Star, *Sorting Things Out: Classification and Its Consequences*

Charles Bazerman, *The Languages of Edison's Light*

Janet Abbate, *Inventing the Internet*

Herbert Gottweis, *Governing Molecules: The Discursive Politics of Genetic Engineering in Europe and the United States*

Kathryn Henderson, *On Line and On Paper: Visual Representation, Visual Culture, and Computer Graphics in Design Engineering*

Susanne K. Schmidt and Raymund Werle, *Coordinating Technology: Studies in the International Standardization of Telecommunications*

Marc Berg, *Rationalizing Medical Work: Decision Support Techniques and Medical Practices*

Eda Kranakis, *Constructing a Bridge: An Exploration of Engineering Culture, Design, and Research in Nineteenth-Century France and America*

Paul N. Edwards, *The Closed World: Computers and the Politics of Discourse in Cold War America*

Donald MacKenzie, *Knowing Machines: Essays on Technical Change*

Wiebe E. Bijker, *Of Bicycles, Bakelites, and Bulbs: Toward a Theory of Sociotechnical Change*

Louis L. Bucciarelli, *Designing Engineers*

Geoffrey C. Bowker, *Science on the Run: Information Management and Industrial Geophysics at Schlumberger, 1920–1940*

Wiebe E. Bijker and John Law, editors, *Shaping Technology / Building Society: Studies in Sociotechnical Change*

Stuart Blume, *Insight and Industry: On the Dynamics of Technological Change in Medicine*

Donald MacKenzie, *Inventing Accuracy: A Historical Sociology of Nuclear Missile Guidance*

Pamela E. Mack, *Viewing the Earth: The Social Construction of the Landsat Satellite System*

H. M. Collins, *Artificial Experts: Social Knowledge and Intelligent Machines*

http://mitpress.mit.edu/books/series/inside-technology

Publisher contact:
The MIT Press
Massachusetts Institute of Technology
77 Massachusetts Avenue, Cambridge, MA 02139
mitpress.mit.edu

EU Authorised Representative:
Easy Access System Europe, Mustamäe tee 50,
10621 Tallinn, Estonia
gpsr.requests@easproject.com

Printed by Integrated Books International,
United States of America